JOHN LEWIS-STEMPEL IS TWICE WINNER OF THE
WAINWRIGHT PRIZE FOR NATURE WRITING.

'*It is a pleasure* to be in the company of a man who is so
attuned to his woody world . . . He is good at sketching
nature, fixing a vivid image in the mind's eye of a reader . . .
Lewis-Stempel has rightly won himself the reputation as being
among our best nature writers . . . *The Wood* is an entertaining,
illuminating, well-turned read.'

Robbie Millen, *The Times*

'A lyrical diary documenting a year in nature . . .
he's *brilliant on birds and their habits.*'

Helen Brown, *Daily Mail*

'This is countryside writing *crackling with vitality*.
I savoured every month spent in this exquisite sylvan zoo.
And — the hallmark of a great read — I learned a lot.'

Nicholas Crane

'*An energetic narrative*, tumbling with facts, judgements
and observations, as well as bursts of humour.'

Ruth Pavey, *Times Literary Supplement*

'Indisputably *one of the greatest nature writers*
of his generation.'

Country Life

www.penguin.co.uk

The Wood

The Life and Times of Cockshutt Wood

John Lewis-Stempel

BLACK SWAN

TRANSWORLD PUBLISHERS
61–63 Uxbridge Road, London W5 5SA
www.penguin.co.uk

Transworld is part of the Penguin Random House group of companies
whose addresses can be found at global.penguinrandomhouse.com

Penguin
Random House
UK

First published in Great Britain in 2018 by Doubleday
an imprint of Transworld Publishers
Black Swan edition published 2019

A CIP catalogue record for this book
is available from the British Library.

ISBN
9781784162436

Typeset in 12/15.9pt Goudy Old Style by Falcon Oast Graphic Art Ltd.
Printed and bound in Great Britain by Clays Ltd, Elcograf S.p.A.

Penguin Random House is committed to a sustainable
future for our business, our readers and our planet. This book
is made from Forest Stewardship Council® certified paper.

MIX
Paper from
responsible sources
FSC® C018179

5 7 9 10 8 6

'You know I am no traveller. I am always wanting to settle down like a tree, for ever.'

Edward Thomas

Preface

This is the story of a wood; its natural daily life, its historical times.

A particular wood, but a wood which can stand as exemplar for all the small woods of England. Ever.

Cockshutt Wood, in south-west Herefordshire, is three and a half acres of mixed (deciduous and coniferous) woodland with a secluded pool where the winter moon lives.

I managed the wood for four years, so I knew it, from the bottom of its beech roots to the tip of its oaks. I knew the animals that lived there – the fox, the pheasants, the wood mice, the tawny owl – and where the best bluebells grew. (A very British quiddity, bluebells in a wood.)

I know why woods like Cockshutt are special: they are, for many fauna and flora, the last refuge. They are fortresses of nature against the tide of people and agribusiness.

Cockshutt was a sanctuary for me too; a place of ceaseless seasonal wonder where I withdrew into

1

tranquillity. No one comes looking for you in a wood. You are safe from prying eyes, just another dark vertical shape among others: a human tree trunk.

In the heart of Cockshutt there was often just the sound of nature; the leather creak of an old oak in spring wind, the drumming of the woodpecker, the lapping tongue of the badger at the pool edge at dawn.

Actually, I lie. Sometimes there was the bark of a domestic pig, and the bronchitic rasp of a saw. I managed Cockshutt in the best way of all, the old way; by letting primitive livestock roam there, and by 'coppicing' it.

Every book needs a justification, and a book about a wood perhaps most of all. How many trees will be felled for its pages? The excuse – no, the *reason* – I proffer is this: a wood should not be a museum. The notion of woodland as static, stately, fixed by ranks of mature trees is modern, and false. In Cockshutt a half-forgotten memory became real. Cows and pigs were raised there, as they had been in medieval days, and the wood provided us with everything from kindling to mushrooms for breakfast. Cockshutt was a working wood.

A wood is different to a forest. A wood is wild, but not so wild it is frightening. You cannot get physically lost in a wood, only spiritually and imaginatively absorbed.

I knew the trees of Cockshutt. Every one of them. A wood is personal; a forest is always remote. There is too much of it. So, I discovered the secret at the heart of Cockshutt, the little stand of wild service trees that are the vestige of the wildwood that was there at the very beginning.

This is the diary of my last year at Cockshutt. I did not go on holiday that year, I did not miss a day of being with my wood.

DECEMBER

A Walk in the Wood

A walk to the far end of Cockshutt – my woodless life –
woodcock – vixen barking – Jew's ears – amid the winter
ruins of the oaks – 'agroforestry' – 'Old Brown', the
tawny – what's in a wood's name? – holly – tending cattle
in the wood – Cold Song – yule log – 'To the British the
oak was as the buffalo to the Sioux'

1 DECEMBER: Into the wood.

Over the stile, on to the path, which runs along
the entire western side of Cockshutt. Past the sweet
chestnut; an amused fingertip greeting to the giant
beech, with its cold slate for bark. To my left the woodland glade we made by hacking away brambles and
sycamore, to my right the narrow dingle where in March
the kingcups bloom yellow, and where in November
the world's mists are manufactured. One giant sycamore remains, pegging down this flappy edge of the
wood.

It is about 3pm; and the rooks are flying home to
St Weonards, not the usual ragged flight but a silent,
determined oaring. As straight as crows.

Past a slumbering ash with a rabbit burrow at the
base, hard pellets at the entrance. Down in the dingle,
which is parallel to the path, the mauve haze of alder
catkins.

On the far side of the dingle, a scattering of mature
ash trees, including one wrapped in ivy; it is an avian

tower block, home to treecreeper, tawny owl and, in the penthouse, wood pigeon. Beyond the ash, the remnant of the old grass 'ride', or trackway, and beyond that, stone barns which once ate into the wood, but are now dead themselves. Beyond the ruins, more ash.

Faster now, to beat the early-falling darkness. How barren, how gloomy the squirrel's drey looks in the wild cherry.

The tracery of the bare trees against the blank winter sky: a sort of scripture. Or an enclosing net.

The wood slowly climbs a bank. Down in the dingle, wrecks of fallen trunk and branches. And alder with their exposed, rat-tail roots.

At the heart of the wood now, which is longer than it is fat, reaching the pool with its ring of reeds. They are grey, spavined; the whole day is grey. A V travels the dull water; the bow wave of the moorhen, who swivels to flash her warning white taillight.

I love the moorhen. Every pond should have a moorhen. It should be the law. In the centre of the pool, which extends to about a quarter of an acre, is a Swallows and Amazons island, with five alder trees. In this late, low light it is a barge under way.

Silver birch line the western bank of the pool; the far side is hazel, alder and sallow.

On, on. I have a personal law of gravity: the faster you walk, the lighter the load. (If Isaac Newton had spent less time loafing under apple trees he might have discovered it too.) Squatted on my shoulder is a bale of hay.

Through the stand of spruce; grim regimented Norwegian sentinels on parade, stuck in their private, perpetual acidic twilight. (The Norwegian spruce, strangers to our shores, were planted on some half-baked subsidy plan in the 1970s, and choked out all the flowers beneath them.)

The dying winter sun, white and nuclear, strobes through the passing larch.

No birds sing, except for a fitful robin in a young beech tree hung with crumpled, copper leaves. In all the wood, this one beech is the sole tree to retain its dress.

December, when the trees stand in their naked truth, is the time to see and assess a wood.

The robin drops his phrase, then restarts, as if my passing by reminded him of his purpose.

Further into the wood: following the faint ink-line of the clay path as it threads by my favourite trees, the Wishbone Oak and the giant Californian redwood ('Hello, Big Boy'). What dream, what hope caused a farmer a century ago on this absolute edge of England, where it runs unseen into Wales, to plant a sequoia?

Nearly dark now: the quarter-moon is failing to break through the cloud that came on the western sky. I have almost reached the grove of kingly oaks, which tower above all other trees in the wood except the sequoia.

The character of trees depends on the season; in spring, they watch you. In early winter, in solitude and great empty skies, they have no more botany than stone.

Tonight, the seven oaks are the temple pillars of a lost civilization.

At the fork in the path I do not haver, I keep left, to year AD 01, or thereabouts, and pass the clump of three wild service trees. The trio of *Sorbus torminalis* are a remnant of the original wildwood. Cockshutt existed when William conquered, it existed when the Romans trod their road to Hereford.

The right fork leads across the neck of the wood, through brambles, to the hollies, the sallow; and the tangles of honeysuckle bines, which, one feels, if tugged would pull down the entire wood. There too is the foxes' den.

My journey's end: the four red poll cows lie in a ragged circle in the last glade, where they watch for the sabre-toothed tigers of bovine nightmares.

I throw the bale in their metal feeder. Beyond the hay rack is a final stand of spruce and larch.

This is Cockshutt Wood: a small wood, chiefly of ash, oak and sallow, in an archipelago of similar small woods in rolling far-west Herefordshire. From above, it would have the shape of a willow leaf, the tip to the north. Around Cockshutt eddy pasture fields, and one small cornfield in my care. Along the western edge is one long field – not farmed by me – put down to a repetitive cycle of wheat and oil rape. Beyond this field, the peregrine-stooped Black Mountains can be gazed. To reach Cockshutt from the lane where I park the Land Rover I cross a paddock, given over to our pigs.

I head back down the path. It is now dark, but it does not matter. I have walked this path so often in the last three years I can walk it in the blindness of night.

I came late to woods. I am not really a woodsman, although my paternal great-grandfather was the 'reeve' (a fantastically ancient appellation), or manager of woodland, for Barts Hospital estate at Aconbury, just over the hill. But then the Great War came and the family were uprooted.

I was born into farming, meaning a woodless childhood, except for the circular hilltop copse at Westhide. As with many scatterings of woodland on farmland, its sole purpose was cover for gamebirds. Kicking through the leaf litter I and my friends sought spent, shiny shotgun cartridges, as colourful as exotic birds (except for the orange Eley cartridges, which no one wanted). Collecting shotgun cartridges passed for a hobby in 1970s Herefordshire.

All my other woodland experiences as a child are mere fragments, bark chippings, of memory:

i) In the attic, my stepmother's dissertation, for her teaching degree, on deciduous trees of Haugh Wood, with examples of buds held to the page by Fablon, a transparent sticky plastic, as certain a signifier of the 1970s as Black Forest gateau and space hoppers.

ii) Picking bluebells in Haugh Wood/picking wild daffodils in Bent Orchard.

iii) There were trees though; the Georgian oaks in the parkland at New Court, Lugwardine, where I learned to ride a horse; the pear tree on the front lawn with its attached treecreeper; our apple orchard; gathering conkers, with their skins as polished as teak, their sour yeasty smell; the tricks and the cheats of playing conkers, such as soaking in vinegar, baking in the oven.

In a small but I think honourable snub to Health and Safety I sent my son to a school where playing conkers was compulsory.

2 DECEMBER: In the shaded places, the frost remains all morning.

What's in a name? Sometimes, there is the archaeology of meaning. 'Cock' is from woodcock, the bird. Shutt is Medieval English for shut or trapped. Cockshutt Wood is where, centuries ago, woodcock were netted.

Once woodcock were common (hence the ubiquity

of Cockshutt as an English wood name). Some still nest locally, over on Ewyas Harold Common, but the four woodcock that clump together for warmth in the bramble of Cockshutt today are autumn migrants. When God made the dumpy woodcock He was in the same whimsical frame of mind as when He cobbled up the platypus. Although the size of a hand, the wood-cock has a stiletto knife stuck on its face.

The bird books label the woodcock's brown-and-white, flecked-and-striped feathers as 'crypsis'; 'magick' would be closer to the mark. Only the curlew, the wryneck and the snipe possess equally effective cam-ouflage. In the thaw, the woodcock's rufous plumage blends it into the leaf litter of the woodland floor.

Woodcock are seamless with their surroundings. They are the leaf blown through the beech grove, the rotted elder stump beside the path, the speck of grey in the shadows of evening.

I know the four woodcock are down in the bramble only because I saw them blasted in by the east wind. Exhausted, they plummeted to earth as dawn rose and I was collecting kindling, and the frost illuminated the stalks of the rosebay willow herb.

A note on English firewood: sappy or dry, ash makes white-hot wood; silver birch is flaming mad, pyro-maniacal; pines are frenetic; apple, cherry and willow sweetly fragrant; holly will burn green, and bright; oak is slow, solid burning, like coal, and similarly acrid.

The saying 'by hook or by crook' comes from the Middle Ages when villagers were allowed to take only dead wood, not cut down trees or bushes. Fallen timber and dead wood could be cleared and pulled out with a shepherd's crook or a weeding hook.

Although people were punished for taking or damaging live trees in medieval England, the law was not as brutal as the Old German laws mentioned by the Roman author Tacitus. He noted that the penalty for someone who dared peel the bark of a living tree (and thus kill the tree) was to have his navel cut out and nailed to the tree and then be driven around the tree until all his guts were wound about its trunk.

John Gilbert, the Bishop of Hereford, settled for excommunicating persons who cut down trees from the nearby Wood of Ross in 1383.

Collecting or cutting firewood produces warmth; burning firewood produces heat. A virtuous circle.

Wood was the first serious fuel. So each day as I bend to pick up ash and willow twigs and branchlets for kindling I am merely a picture echo of Stone Age man. We humans have been utilizing the wildwood for warmth for five hundred thousand years.

3 DECEMBER: I set up a white plastic garden chair by the pool, at the line where the deciduous trees give way to the pines; the middle of Cockshutt, where I can see the wood before and aft. The view from this seat will be time-lapse photography in my memory.

The stand of spruce: viewed from a certain angle, the spruce are so close together they form an impassable wooden wall. The smell of firs varies; sometimes it fills the air, sometimes, like today (refrigerated and dry), it is non-existent.

To my front, the poolside birch, the snow queen trees; no one, ever, thought of silver birch as male. Despite its gracefulness, birch is tough; it was the first tree to colonize Britain after the last Ice Age.

Silver birch, *Betula pendula*, is distinct from its cousin the downy birch, which does not have bark bosses at its base.

How our woods were made: when the ice retreated ten thousand years ago Britain was treeless tundra joined to the Continent. As the climate improved, the time was ripe for arboreal invasion. After the birch came juniper, willow and Scots pine, to form the advance guard of trees.

In the corner of my eye, movement. The vixen. In the half-light her fur lustres. She burns alive.

She screams her mating call, which is the wail of the bereaved.

If I had been cold before, I am colder now. She

howls again, and cocks her ears for a response. Nothing comes back over the bitter-hard fields or through the trees. She trots on, is almost in front of me when she finally catches my scent. A flash of her face, then she bolts and is gone, an extinguished flame.

In a wood, the trees are the star turns; but I do applaud the support acts.

5 DECEMBER: Fog huddles down in the dingle; along the bottom is a trickle-stream from the pool. A heron comes up through the dingle fog, exactly in the style of an Ancient Greek trireme riding surf, to berth at the side of the pool.

The bracket fungus on the willow is a dry shelf; when tapped it sounds like polystyrene. In a life form unassuming, inert, is held, perhaps, our human future; bracket fungi could be the next source of antibiotics.

The woodcock have flown.

Sometimes I worry that Cockshutt is so small that it is, properly, a copse. But it feels old, and from the lane there is the impression of solidity, of belonging.

In the beginning was the wildwood. Then came the Stone Age people, who made the first holes in the limitless canopy. The wildwood provided nomadic hunters with game, berries, edible roots, leaves, seeds, cut wood, and covered almost the entire land of Britain, including the hilltops. Pollen counts from prehistoric peat show a sudden decline in elm from about 3000 BC, and a corresponding increase in nettles. The dismantling of the wildwood by Neolithic farmers had begun; the Stone Agers killed most of their selected trees by ring-barking, chipping a horizontal skin-wound with a stone axe, in imitation of the deer. Then came the Celts with their iron tools. The Romans turned much of lowland Britain into an imperial bread basket; yet in the ninth century large tracts of natural forest remained. The Kent and Sussex Weald stretched for 120 miles. It was the agriculturalist Anglo-Saxons who transformed the landscape; they fixed boundaries of field, woods and parishes still extant today. By the time of the Norman Conquest, the open pattern of our modern countryside was established. Only some 15 per cent of the land recorded by the Domesday survey was woodland and wood pasture. It is thought that the Forest of Dean was the last natural wildwood in England, and that was felled in the thirteenth century.

We managed our woods back then. The products of coppiced woodland warmed village and city hearths alike. (To coppice, from the French *couper*, means to cut down to the ground, so the trees regrow. The coppice cycle varies with the species; hazel is cut on a seven- to twelve-year cycle, ash every twelve to fifteen years,

17

and oak every thirty years or so. Pollards are up-in-the-air coppice, where the trees are cut above the browsing reach of livestock.) The woods provided the timber for the British battle fleets of Cromwell and George III that won an empire. Our native woodlands fell into disuse as man's endeavours turned in other directions with the Industrial Revolution, fuelled by coal and producing iron.

Two world wars, the felling of timber to pay death duties and the scourge of Dutch elm disease carried off much of what once remained.

Only thirty-nine trees are believed to be native. Of these Cockshutt has:

oak	larch
beech	alder
sallow	elder
hazel	elm
wild cherry	service
silver birch	holly
ash	

Cockshutt's 'alien', imported trees are spruce, Canadian redwood, sweet chestnut, sycamore.

And the difference between trees and shrubs? Shrubs have many 'sticks', trees a single trunk. Generally.

6 DECEMBER: A milky-watery sultriness over everything. Windless. Almost a prelude to spring.

I let the piglets out of the paddock at the bottom of Cockshutt into the dingle; they race around, the nine of them, as one thing.

They were born only a month ago; elf-eared, ribby, plunger-mouthed fresh-born piglets are unattractive. They have become cuter with age.

In the unseasonal mildness an eruption of conical *Cononcybe tenera* mushrooms near the pond; a rotting sci-fi city.

Later, close to nightfall: a jackdaw murmuration above the wood. The murmuration clenches to a ball, then opens in the shape of a lady's fan, becomes a Chinese dragon a hundred feet long, a rug shaken by unseen hands.

Woods have ears. The recent mild, moist days have caused the biggest crop of Jew's ear mushrooms on the lolling elder tree in the dingle I have ever seen. The Jew's ear fungus derives its name from its extraordinary resemblance to the human hearing apparatus, together with the ancient belief that Judas Iscariot, having betrayed Christ for thirty pieces of silver, hanged himself from an elder tree. (The Latin name is *Auricularia auricula-judae*, meaning 'ear of Judas', corrupted over time to Jew's ear.) Local names include wood ear and jelly ear, the latter recognizing the gelatinous, rubbery texture of the fungi when tumescent.

One of the beauties of the ugly Jew's ear is that it is impossible to mistake for any poisonous fungus. I take one of the Jew's ears home; it measures five inches wide. Good enough to tweet.

3.30pm. Redwings land in the hazels behind the pond: a waterfall of chattering.

4.30pm. Red sky lit by the fires of Hephaestus. Gold-crests tease out seeds in the alder's small black cones.

Another jackdaw murmuration: two hundred birds this time, with small bands joining the mass, as tributaries join a river in confluence. The jackdaws *jack* loudest when changing direction.

From the far side of the wood: *Hoo-hoo . . . H-o-o-o-o-o-o-o*. Night music: the hoot of a tawny.

Is there an index of owl happiness? I think so. For the three years we have been managing the wood Old Brown's wives have produced steadily bigger clutches. Two eggs. Three eggs. And this year, four eggs.

The reason for the increasing size of the clutches is that we have improved the tawnies' food supply by reducing the invasive bramble, nature's barbed wire, which covered almost the entire woodland floor. Before Old Brown was simply unable to penetrate the bramble when out a-hunting.

I say 'we' have reduced the invasive, hegemonic bramble, but the real graft has been done by the beasts, by the hooves and the teeth of the cows, pigs and sheep. More than a third of the woodland floor now is leaf litter, fallen boughs and grassy glades. No longer does Old Brown live on the margins, or need to fly from wood to wood. His meals of shrew and mice scurry in their hundreds around his home.

8 December: I shoot a pheasant (for Christmas lunch), under a black cloud the size of Europe.

The pheasant, a cock, in his usual place, the coppiced alder at the corner of the pool. He lands with a thump. I regret killing him instantly. I liked his presence in the fields and wood.

The side-splashes on his face are scarlet to remind me I have spilt his blood.

The afternoon light does not fade, it sinks, stone-speed.

Pheasants have probably been here since Roman times. Palladius, writing in about AD 350, advises on how to rear them. Definite proof of their existence is provided in a Waltham Abbey Ordinance of 1059. The pheasant has always been preserved, fiercely so, by the shooting fraternity. On 18 January 1816 one John Allen and a group of men from Gloucestershire mounted a poaching expedition on the estate of Colonel Berkeley at Berkeley Castle. Berkeley's keeper, Thomas Clarke, and nine other keepers were waiting for them. Someone fired a gun (allegedly the poacher John Penny), killing a keeper. The poachers were tried on 3 April 1816 for murder. John Allen and John Penny were sentenced to death, the rest to transportation.

About 25 million pheasants are reared and released annually in the UK. The pheasant in the alder was a wanderer from a shoot, although pheasants do breed in Cockshutt.

9 DECEMBER: The sky is bruised mauve.

It's as I turn to go to the Land Rover, parked up on the lane, after dropping off the hay to the cows in the oaks, that it starts snowing. The snow is harsh, and shimmers noisily.

And I think: being in a wood on a winter's eve, when the snow is falling through leafless oaks, is existence stripped back to the elements.

10 DECEMBER: Snow is not usual in December despite the wants of my children, but this is high country so we do sometimes get it.

Snow makes everything old, including us, who stoop before it.

Fallen snow paralyses sound. The silence of high

summer is actually a background hum of mono-tonous similar sounds: bee-drone; woodpigeon-coo; grasshopper-zizz.

Standing in the wood, among the black trees, the blue afternoon snow at my feet, looking out.

Beside me a solitary treecreeper searches an alder's bark, in the same way a janitor checks under the audi-torium seats for rubbish after a concert.

The snow goes quickly, except under the hedges, where it lies in scars for days.

Stopping by Woods on a Snowy Evening

Whose woods these are I think I know.
His house is in the village though;
He will not see me stopping here
To watch his woods fill up with snow.
My little horse must think it queer
To stop without a farmhouse near
Between the woods and frozen lake
The darkest evening of the year.
He gives his harness bells a shake
To ask if there is some mistake.
The only other sound's the sweep
Of easy wind and downy flake.
The woods are lovely, dark and deep.
But I have promises to keep,

> *And miles to go before I sleep,*
> *And miles to go before I sleep.*
>
> Robert Frost

12 DECEMBER: A sorcerer's dawn, red and violent; an ember of kestrel sits on the telephone wire.

The domestic pig is not an early riser. I'm taking breakfast in bed to the sows and nine piglets in the paddock; against type the piglets have got up early, and escaped out of the night pen into the wood.

I run through the trees, impenetrable at a distance, but close up they step aside for me. Finally I find the scampy piglets and shoo them back for their breakfast. Later in the morning, back at home, a customer phones and asks: 'Are your pigs free range?' Me: 'Madam, they are practically wild.'

The afternoon is all mizzle and mist; my breath adds to the white veil. An unseen squirrel scolds. The pond trembles with tiny raindrops; the woodland floor has no colour, except bark brown. All the nettles and the sky-scraping rosebay willow herb have now collapsed under the weight of the season, the unsloughable burden of time.

7.25pm. Arrive at the lane. I can hear a dog fox barking up in Cockshutt; it is posh-fox barking. *Rah. Rah. Rah.*

14 DECEMBER: The rabbits have scratched at a pollarded alder, causing pink sores on its thrashing, raised roots.

In the sunshine the Norwegian spruces are colourful with their bright green needles and long orange armadillo cones, I give them that.

Today, I finished a job I started three years ago – a cadastral survey of all trees in the wood. There are 647 trees above the rank of sapling.

A note on terminology: in modern English forest and woodland are used interchangeably, except that forest suggests a larger area of trees, for example New Forest, Sherwood Forest. 'Twas not thus in the past.

Like the French *forêt*, the English word forest has at its root the Medieval Latin *foresta*, in turn probably derived from late-Latin *foras* meaning 'outside' ordinary jurisdiction, being subject to separate 'forest law'. Forest law was primarily designed to protect and provide game for the king's table. A forest would include large areas of land that were treeless, such as farmland, and even whole towns.

We have lost the clear definition of forest in English. An area of trees can be a forest or a wood, though forest has come to suggest super-size. Woodland is a term associated with naturalness, but natural woodland is a misconception. Along with coppicing and wood-cutting, from the Bronze Age onwards woods were farmed with livestock.

This knowledge has all but lapsed from Western memory, although trees and flower names preserve it. Sallow is also 'goat willow', because goats will readily browse on its leaves.

All common livestock, except sheep, are descended from woodland beasts. Much of the medieval economy depended on marauding pigs in woodland. Chickens prefer woodland; they are, after all, descended from Indian jungle fowl. Aerial predators need a clear 'run' – trees stop this. Free range chickens are reluctant to go outside into treeless spaces. For cattle, pigs and chickens, woodland provides tubers, nuts, invertebrates (for swine and fowl), along with 'browse' (herbage, foliage), and shelter, shade. In return, the livestock and fowl manure the land, knock bits of branch off, open up dense undergrowth to light. The rootling of pigs and chickens, the churning of cow hooves provide ideal conditions for the germination of seeds and nuts. The beasts imitate the actions of the aurochs and the wild boar which once roamed the wildwood. Save for deer, the influence of larger animals on woodland was largely lost between the later Middle Ages and the twenty-first century.

I am not the only one who now runs livestock in woodland. The new-fangled farming in woodland even has a name, 'agroforestry', meaning the production of trees alongside the production of livestock or conventional food crops.

15 DECEMBER: An ice moon in the morning, perfectly round, perfectly white.

The heron stands in the dingle; after the rain and the melt the dingle bottom is an African delta seen from an aeroplane; shining silver, branched waters.

Robin song trickles through the wood. A grey squirrel jumps up a birch – emphatically he does not climb, but lurches up and grabs: repeat – then goes over the canopy above my chair; one branch too slender, tries another – continues his parkour performance before climbing down and walking along, insouciantly.

There is always the sense of the unexpected in a wood, a constant feeling that, around the next bend in the path, behind the bole of the next tree, there will be a surprise.

16 DECEMBER: Dusk: the wood pigeons fly into the larch, in ones and twos, till about thirty settle.

At the impenetrable pond; the reeds dissolve out of existence into night.

A surprising amount of birdsong: blackbird, pheasant, wren.

Woods: they inhabit the mind.

Old Brown the tawny starts his hooting; he is announcing his rule of this wood. Tawnies are most vocal in autumn, when the male hoots in claim of breeding territory. The first call is heard in late September once the birds have completed their moult; the calls increase in frequency until December. Creatures of the night, owls communicate primarily by sound.

The Saxons knew the bird as *Ule*, named after its lamenting cry. The tawny is Shakespeare's 'clamorous owl, that nightly hoots and wonders at our quaint spirits'.

Old Brown will keep calling till about 7.45 tomorrow morning, a long shift. The owl that sings loudest, longest is likely to be of better stock than one who can only afford to sing for short periods.

It is three years ago that Old Brown won the territorial rights to Cockshutt. He has since held it against all comers – owls, foxes, badgers. He once had a go at our Jack Russell; luckily I was on hand.

I link my fingers together, then press the palms together and blow a facsimile hoot to Old Brown.

It takes three attempts, but then he replies. *Hoo-hoo-hoo-h-o-o-o*. What a hoot.

Collective nouns for trees:
avenue – a line of trees, one or more rows deep, each side of a road

brake – a clump of shrubs, brushwood, briars or fallen trees (cf. thicket)

coombe – the head of a wood in a valley

coppice – an area of woodland where the shrubs (e.g. hazel) are cut regularly for wood or 'browse', fodder for livestock

copse – a small woodland, a half-acre or less

covert – a dense group of trees or shrubs, often connected with game rearing and shooting

dingle – a deep wooded valley

grove – a small group of trees without undergrowth

hagg – a small group of trees (cf. stand)

hanger – a wood at the top of a rise, from the OE *hangra*, wooded slope

plantation – an area of artificial woodland, composed usually of conifers

spinney – a copse, often of thorny shrubs and trees, that shelters game or foxes

stand – a small group of trees. Also used by foresters to describe a particular group of trees under similar management

thicket – a dense growth of shrubs and briars

wood – used interchangeably with woodland, for an area of tree-covered land bigger than a copse but smaller than a forest

17 DECEMBER: On the far bank of the pond the heron stands in the mess of dead crossed-sword reeds, between two alder pillars, looking fiercely pharisaical in the way that herons do. The temple priest of the pond, deciding I am too close for safety, flaps off on slow, sad wings. He emits a single *craaak*, which cracks the valley's frozen silence apart.

I am sorry to have disturbed his fishing.

There is always a certain amount of Mr Bean comedy in cutting holly with pruners. Sure enough, as I am climbing up the ladder a bough springs back, catching my forehead. I bleed berries.

But I have to go up the ladder because of 'spinescence'. Holly leaves at the bottom of the tree are armed with prickles to deter grazing animals. Leaves above grazing height are spikeless ovals, as smooth as on a camellia.

Centuries ago someone planted a 'hagg', or stand, of holly in the wood, probably to provide winter fodder. When the hay ran out, or when it snowed, the upper branches of the *Ilex aquifolium* were lopped off to feed cloven-hooved stock. There are five coppiced holly trees remaining. A living remnant of the old farming ways, they are as tall and green as firs.

Using holly as fodder is one of those agricultural practices that, along with scything and the making of corn dollies, has gone to the rural museum. But the custom of feeding holly was of great importance in the years before hay and turnip winter feed, especially in the west and in the hills. The deliberately planted holly grove was important enough to warrant a name, hollin, as in the Hollingwood Grange in the Golden Valley, just a few miles up from us. Like holly itself, 'hollin' place-names in England are concentrated in the north and north Midlands, away from the lowlands and more fertile soils.

There is a scene in 'The Dream of Rhonabwy', a Medieval Welsh tale from the twelfth century, where the quarters for the livestock have 'branches of holly a-plenty on the floor after the cattle had eaten off their tips'. Holly was also used as fodder for the lord of the manor's deer. Illegal cutting was punishable by the bleeding of your wallet. At Tideswell Court in the Royal Forest of the High Peak ten people were fined for lopping 'green-wood' in 1524, twenty-four were subsequently fined in 1559, and twenty-one in 1567. Holly was green money; rents for holly groves could be as much as £1 16s 0d a year in the eighteenth century. William Cobbett, during his younger years working at Farnham in the 1770s, apparently spent time grinding holly to make it more acceptable to domestic stock. Up on Exmoor Richard Jefferies, the Victorian nature writer, noted, 'Wire fencing has been put round many of the hollies to protect them' from deer.

By then, the practice of using holly as fodder was dying out. Also the use of holly trees as a source of

shelter for livestock became superfluous with the building of dry-stone walls on the enclosed commons.

I use holly as a pick-me-up or vitamin supplement
for cows and sheep. You have your echinacea and white
chia – the moos and the ewes have holly.

It was my grandfather who taught me to use the
plants as medicine or tonic. Whenever any farm animal
or horse was 'not looking too clever', which is what we
say in Herefordshire for sick, Joe Amos would let said
animals browse the hedges to cure themselves. Just over
the mountain, Welsh farms used to have a *cae ysbyty*
or 'hospital field', with healing wild flowers and plants.

After four hours of cutting I have a heap of glossy
holly leaves stacked on a tarpaulin which I drag up to
the cows.

I thread about half the holly into the wire stock
fence alongside the north-east edge of the wood; a
woven fence of Christmas decorations for livestock,
and a welcome wall of green in a leaden landscape.

One of the cows ambles over, inquisitive, wraps her
pink tongue around a holly cutting and starts chewing.
Another joins her. They stare into the distance, across
the quiet winter fields, as if they can see some secret I
cannot.

18 DECEMBER: 3pm. The trees fret in the coming
Siberian blast.

We no longer fear winter like people did, when their clothes were thin and ragged and they were obliged to be out in the woods, tending cattle as I am today. Shakespeare mentions it over and over. In *King Lear*, there is that haunting line of Edgar: 'Still through the hawthorn blows the cold wind.'

People claim they enjoy winter, but what they really mean is they enjoy winter as a livener, a cobweb-blower-away, a quick flirt with the elements before resorting to their real love, central heating.

For anyone working outdoors, like I am, winter hurts. The bale of hay on my shoulder scratches into my neck; the baler twine slices through my glove into my left hand. So does the wind, which comes seeking down the path, seeking me.

I cannot shake out of my head the 'Cold Song' libretto Dryden wrote for Henry Purcell:

> *See'st thou not how stiff and wondrous old,*
> *Far unfit to bear the bitter cold,*
> *I can scarcely move or draw my breath?*
> *Let me, let me freeze again to death.*

The four red poll cows are at the top of the wood, in the middle of the oaks, forlorn backs to the wind. I toss the hay bale into the rack and almost run back to the Land Rover, to throw myself on the mercy of the heater.

19 DECEMBER: I decide to let all the pigs out of the paddock to have a rummage at the very south end of the wood, where they bulldozer away all morning, particularly under the sweet chestnut and beech, looking for old mast and new bluebells.

'Mast' in botany is derived from the early German *maesten*, meaning to fatten or feed. A mast year is a natural phenomenon in which certain trees, such as ash, chestnut, English oak and beech, produce a glut of seeds. Beech, for example, produce a mast year every five to ten years. No one knows what produces a mast year, though self-protection, through fecundity, would seem to be an obvious explanation.

There is a surprise in the snout-ploughed earth. A rusted Victorian gin-trap. This wood could tell tales.

Often oaks will keep their leaves until February, but not this year. From one ivy-clad ruin a wren, as small as a moth, peers at me. It is too feeble to *tisk* its default alarm.

The ground is iron, and the previous night's frost has frozen-in the usual toadstool smell of woodland. There is just the exhilarating purity of ice, a wipe-clean of the senses. But a rope of rank scent stretches across the path by the stile – the odour of fox.

The winter wood echoes with sounds distinct to itself: the hand-clap of pigeon wings as they exit the skeletal trees, the paper-rustle of rabbits scuttling across dry sycamore leaves to hiding.

Deep into the wood I startle a pheasant under briars. Or rather it startles me, racing headlong for take-off. Beak-shriek of pheasant; another sound loving a wood. The bird's feathers, caught in the bramble's barbs, flutter up.

The grey squirrels have entered the mating season, so I blast the two dreys in the wood with the 12-bore, a smoking barrel apiece. The twiggy domes, sequestered unlawfully from a crow, are atomized.

The squirrels are not at home.

In our first year at Cockshutt the rampant squirrels ate every clutch of greater spotted woodpecker eggs, every clutch of blackcap eggs, the chicks of the willow warbler. Call me a nature boy but I rather like songbirds in an English wood.

Lights Out

I have come to the borders of sleep,
The unfathomable deep
Forest where all must lose
Their way, however straight,
Or winding, soon or late;
They cannot choose.

The Wood

Many a road and track
That, since the dawn's first crack,
Up to the forest brink,
Deceived the travellers,
Suddenly now blurs,
And in they sink.

Here love ends,
Despair, ambition ends,
All pleasure and all trouble,
Although most sweet or bitter,
Here ends in sleep that is sweeter
Than tasks most noble.

There is not any book
Or face of dearest look
That I would not turn from now
To go into the unknown
I must enter and leave alone,
I know not how.

The tall forest towers;
Its cloudy foliage lowers
Ahead, shelf above shelf;
Its silence I hear and obey
That I may lose my way
And myself.

                                        Edward Thomas

20 DECEMBER: Our pigs are Large Blacks, Welsh, Berkshire: native breeds, with some bristle on their backs to protect against the elements, hardy character, long noses for rummaging. In a wood they forage – saving me money – and improve their health. Pigs are what they eat.

Alas, pigs are indiscriminate feeders, meaning they will happily plough up the two grassy glades I have made by clear felling. Neither are saplings safe from the pigs' sharky mouths. Moving the pigs around the wood requires electric fences (three horizontal strands of polywire and the big ol' tractor battery, producing a zillion watts to power the deterrent shock) and tree guards with a ring of barbed wire.

Our native swine also have the instinct to get out and about, beyond manmade, arbitrary boundaries. And so the pigs escaped (again) this morning, requiring me to stand on the field gate at the entrance to the ride to prevent Tinkerbell, a 300kg Large Black, getting free and joining her raft of piglets in a neighbour's field, where they spun around happily.

All this makes me late feeding the four red cattle at the top end of the wood; it is about 5pm on this dark night.

Over my head, a slight misplaced breeze. Old Brown is on his rounds.

I wonder, can owls see in the dark? Almost. When it is bat-black like tonight the iris of an owl's eye opens almost completely to allow in all the light there is. Owls have the best 'stereoscopic' vision of all the birds.

With this light-sensitive optical equipment, Old Brown can navigate Cockshutt at night, although he is also dependent on my technique for traversing the wood in the dark: the map in my head.

He screams at the night. For all of our street-lamp civilization, in a December wood at five o'clock, you can still hear the call of the wild.

From a distant farmhouse chimney comes wood-smoke, which is the scent of the countryside in winter. Someone has lit a fire.

21 DECEMBER: 7.38am. Sitting in the silly white plastic poolside chair. Owls still hooting; jackdaws start singing, though more like the warble you get trying to tune a radio than chansoning.

In the afternoon the sun makes the trees into steel engravings; at night the ash grasps the moon.

Few trees hold the moon as well as the ash.

22 DECEMBER: The bleak trees, the lonely silence, the dead leaves underfoot. There are no flowers; it is really flowerlessness that gives December its distinction.

23 DECEMBER: Stormy, wet, windy. The old trees moan, as if they were old people protesting against the indignity of movement.

It always gets dark in a wood first.

24 DECEMBER: It is a year of holly berry dearth. The Viking birds, the redwings and the fieldfares, have descended from the north and plundered all the other hollies for miles around. In just two or three days.

The scarlet and the green of the holly is shockingly vivid on this midwinter afternoon. As Henry VIII wrote in song:

> *A! the holyt grouth grrene*
> *With ive all alone,*
> *When flowerys can not be sene,*
> *And grene wode levys be gone.*
> *Grene growth the holy, so doth the ive:*
> *Thow wynter blastys blow never so bye,*
> *Grene growth the holy.*

Holly is apotropaic, the symbol of Christ, and was for centuries the physical mnemonic to remind us that 'Mary bore sweet Jesus Christ on Christmas day in the morn' because its berries are red like His blood, its prickles ('sharp as any thorn') are akin to the Messiah's crown at crucifixion, and its leaves evergreen, an arboreal metaphor for eternal life.

Is the Nativity superstition? Perhaps. In my family we have always believed it. In late summer my family decked their house with hops, at yuletide with the holly and the ivy.

As a son of the country, I do not care to be without holly at Christmas. As a boy my grandfather gathered holly for the Christmas decorations. On this Christmas Eve I am his reproduction.

In a wood, time obeys different laws. It is always the past.

25 DECEMBER: A log on the fire; an atavistic delight, which takes one way beyond childhood, Victorian coaching inns, Henry VIII's Merrie England, to the beginning of human time.

I myself sawed this log from a fallen arm of oak.

The burning log warms body, soul, the night. And with the curtains drawn, we four are cave people, the walls flickering in firelight's glow – the released sunlight of years gone by.

Historically the yule log burned at Christmas was an oak log, and the remains of the log were kept as charms against fire and lightning, and as kindling to light the next year's yule log. The poet Robert Herrick gives the particulars in his 'Ceremonies for Christmasse', 1638:

Come, bring with a noise,
My merrie merrie boyes,
The Christmas Log to the firing;
While my good Dame, she
Bids ye all be free;
And drink to your hearts desiring.

With the last yeeres brand
Light the new block, and
For good successe in his spending,
On your Psalteries play,
That sweet luck may
Come while the Log is a-tinding.

The anthropologist James George Frazer suggests that this may be a relic of ancient oak-tree worship. The early Celts titled their priests after the oak. Druid derives from *dru* plus *uid*, meaning someone who possessed 'knowledge of the oak'.

Britain has two oaks; the common, pedunculate or English oak, *Quercus robur*, and the sessile oak, *Quercus petraea*. Sessile means sitting, and refers to the way the cups of the acorn sit directly on the twig and do not have a stalk of their own. The oaks in Cockshutt are *Quercus robur*, and have the sort of acorns on long stalks that gnomes use for tobacco pipes.

*

The oak. To the British the oak was as the buffalo to the Sioux. The all-provider. In his *Sylva* (1664), John Evelyn, diarist and xylophile, listed some of the oak's usages in traditional medicine:

> *Young red oaken leaves decocted in wine make an excellent gargle for a sore mouth; and almost every part of this tree is soveraign against fluxes in general, and where astringents are proper. The dew that impearls the leaves in May, insolated, meteorizes and sends up a liquor, which is of admirable effect in ruptures . . . also stops a diarrhoea. And a water distill'd from the acorns is good against the pthisick, stitch in the side, and heals inward ulcers . . . and refrigerates inflammations, being applied with linnen dipp'd therein: nay, the acorns themselves eaten fasting, kill the worms, provoke urine, and (some affirm) break even the stone itself. The coals of oak beaten and mingled with honey cures the carbuncle; to say nothing of the viscus's, polypods, and other excrescences, of which innumerable remedies are composed, noble antidotes, syrups, &c. Nay, 'tis reported, that the very shade of this tree is so wholesome, that the sleeping, or lying under it becomes a present remedy to paralyticks, and recovers those whom the mistaken malign influence of the walnut-tree has smitten.*

Oak gall, crushed and left in water with rusty iron, made a kind of ink. Tannic acid from oak bark was used for tanning leather, while coopers valued the oak for cask-making. Acorns fattened food for pigs. Oak leaves made wine. Oak flavoured whisky. Village match-makers insisted on the efficacy of May dew gathered from oak leaves as a beauty treatment for young women.

Of course, the main reason for the veneration of the oak tree was the timber, which was probably the strongest wood in Britain. The seventeenth-century poet Renatus Rapinus versed:

> *When ships for bloody combat we prepare,*
> *Oak affords plank, and arms our men of war;*
> *Maintains our fires, makes ploughs to till the ground,*
> *For use no timber like the oak is found.*

It is estimated that it took two thousand trees to make one warship in Nelson's navy.

For centuries English churchgoers took advantage of the broad canopies of oaks to mark parish boundaries; these trees became known as gospel oaks, since Rogation processions paused beneath them while a passage was read aloud from the Gospels.

Oaks define the landscape, and our history. It is the prevalence of *Quercus robur* in Britain, more so than in any other Western European country, that forms this pride in what may be seen as our national tree.

At Boscobel in Shropshire stands the Royal Oak, where the would-be King Charles II hid from Cromwell's men after the battle of Worcester in 1651. Robin Hood and his Merry Men are said to have feasted beneath the Major Oak in Sherwood Forest. Oak provided the cruck frames for the black-and-white Tudor cottages of national longing. Oaken ships defeated the Armada, bested the French at Trafalgar. Admiral Lord Collingwood, one of Nelson's band of brothers, walked the hills and lanes of Northumberland with pocketfuls of acorns, planting them in hedgerows and patches of waste ground so that the Royal Navy would never lack oak.

'Hearts of oak are our ships, jolly tars are our men,' we sang at school in the 1970s, the wood-panelled hall hung with Union flags on Remembrance Day. (Not 'Union Jacks'; we were not at sea.)

'Heart of Oak' is the official march of the Royal Navy of the United Kingdom. The music of 'Heart of Oak' was composed by William Boyce, and the words were written by the eighteenth-century English actor David Garrick.

The oak is also recorded in many British proverbs, such as:

An oak is not felled in one stroke – signifying patience.
Great oaks from little acorns grow – from little things grow great things.
The willow will buy a horse before the oak will buy the

saddle – referring to time, as oaks grow much more slowly than willow.

The oak is recognized in folklore for weather-forecasting:

> *Oak before ash,*
> *In for a splash.*
> *Ash before oak,*
> *In for a soak.*

(But surely ash never comes into leaf before oak?)

Another bit of old verse, about safety during a thunderstorm, is:

> *Beware of the oak, as it draws the stroke,*
> *And avoid the ash as it counts the flash.*
> *Best creep under the thorn, as it will keep you from*
> *   harm.*

Early humans likely believed that the lightning hitting oaks and setting them afire was sent by sky gods. From this belief it was a short step to perceiving the oak as sacred, the fire-maker. Some stands of oak are believed to have been the sites of pagan temples.

The Green Man, the primitive's tree spirit often pictured with wise eyes peering from a face composed of oak leaves, was absorbed by Christianity, and can be seen in many old English churches. He is also known as Jack-o'-the-woods.

As the naturalist Brian Vesey-Fitzgerald observed, in the past 'the oak may be said to have touched the lives of Englishmen at every point from cradle to coffin'. Not now, thanks to Ikea fibre-board furniture, steel, and my plastic poolside chair.

27 DECEMBER: The morning is frosty, and at home the redwings in the paddock glow in the dawn.

When I go to the wood in the afternoon, the pond is bare steel, but in a while the reeds will erupt in their accustomed place. The primroses will bloom on the bank.

I spend the afternoon fencing the top of the wood, four horizontal strands of barbed wire against posts, quick and cheap, until dusk. Pheasants *cok-cok* from wood to wood, a chain reaction. The cry of the cock pheasant is another declaration of ownership, the announcing of a fiefdom.

On the way back down the path, I pass under the arm of the Tall Oak, on which is perched a roosting pheasant, who thins and crouches, and watches me warily, as well he might.

As I leave the wood, the jackdaws start up, the original neighbours from hell.

28 December: At the pig paddock I leave the tractor engine running, because on an afternoon like this, when the wind is flailing wire, it is warmth and heartbeat.

Pavlov had a whistle to summon his dogs. I have a half-brick to call our pigs to dinner. I rap the brick bit against a steel trough. *Ding, ding* tings out over the glassland.

The pigs are off in the wood, rummaging. Pannage, the practice of releasing pigs in woodland, was anciently important. Indeed, according to the Herefordshire Domesday not much else mattered locally. 'There was woodland there for 160 pigs, if it had borne mast,' runs the entry for Pembridge. My mother's family held the pannage rights in the Golden Valley, up the lane, till the 1600s, a matter of carefully preserved, and jealously guarded, record.

Pannage also kept beech nuts and acorns away from cattle and horses, who find them poisonous.

The 'girls' are safe enough out in the wood but, akin to an anxious father, I like to make sure they are back in their home field at night.

There is no answer from the pigs. Damn. This time I bash the trough with the brick. *Dong, dong* bells out; the vandal sound echoes around the hard hills.

From the recess of the wood, from the recess of time, there is a pig's answering squeal.

Down come the pigs, shadow shapes weaving in and out of the dulled columns of beech.

The pigs follow the old path, pounded into the clay by generations of animals' feet, by Brock, by Reynard, by Bambi. The path only seems to meander aimlessly. It took me a year of walking it to realize that in its avoidance of subtle hazards and inconveniences it was the fastest, surest way through the trees. Once, some American academics pitched a computer against cows in a competition as to which could find the quickest route over rough terrain. The cattle won every time. You see, the animals know best.

Running pigs are not gainly. They galumph. Neither are they quiet. There is grunting, and as they near me there is the flesh-slap of Dumbo ears. (If these pigs' ears were any bigger, pigs might fly.)

The pigs oink up at me as they come through the gate, a sort of salute, before advancing on the sow breeder nuts in the trough. Pigs eat like pigs, with their mouths open; their slobber is sea-foam phosphorescent in the dusk.

Large Blacks, Berkshires, Welsh, they all get a bit of a rub behind the ear. People say they don't like pigs, but are not sure why. I can tell them. The naked skin of a pig is that of a human. I lift up Lavender's floppy-loppy ears to say hello. She has a girl's eyes.

Lavender, as she well knows, is my porcine darling, my Miss Piggy, but I do not think it blind favouritism to say she wears a lovely perfume. She smells of freshly ironed linen.

The pigs seem particularly bonny tonight. Does landscape change an animal's mood, their behaviour? Pigs are the descendants from the wild swine that inhabited the larger forests of Britain. I do not believe it fanciful that our pigs have the happiness of feeling at home, of existing in their rightful habitat.

29 DECEMBER: Morning: up a ladder, putting up nest boxes on larch and oak, for tits and, more in hope than expectation, flycatchers.

Freezing fog rolls in from the west, adding to the addictive privacy of the wood. I am marooned in my own world, me, alone with the animals, the birds, the trees.

Two cock pheasants in the glade start fighting, head-bobbing, samurai-bowing, plunging at each other, then take a long walk out through the fence into the field, always a train track apart until they drown in the fog. The new pheasant, a big cock with a white collar, has escorted the Tall Oak pheasant away.

The Jew's ears are now shrivelled into scraps of leather parchment.

A rush of jackdaws overhead: hail against the window, the breaking of a wave on the shore.

The pigs eating in cold weather: steam-engine breath, piston-regular, white.

30 DECEMBER: Oddly aware, walking through the wood this afternoon, that it is dormant rather than dead. How the seeds, the trees and hibernating animals (the hedgehog, toad, frog, snake, the thousand butterflies, the billion insects) are locked in a safe sleep against the cold and wet. Under the ground and in hidden corners they wait, for the signal to arise again. In this wood of dark oak, ash, beech, sallow, pine, runs the eternal circle of Zarathustra:

> O man, take care!
> What does the deep midnight declare?
> 'I was asleep –
> From a deep dream I woke and swear:
> The world is deep,
> Deeper than day had been aware.
> Deep is its woe –
> Joy – deeper yet than agony:
> Woe implores: Go!
> But all joy wants eternity –
> Wants deep, wants deep eternity.'

And I do notice that the velvet black buds on the ash are fully formed.

31 DECEMBER: Evening: light and dark dispute longer. The wood is always beguiling. Always. Even on

this minimalist evening, when the air is wetly soft, and milk-toned. Certainly, there are no birds singing prettily, or sweet little flowers . . . but there is the permanent moonlight of silver birch bark, the luminescent halo around the squirrel's brush, the startling cold pewter of beech bole.

There are nine mature beeches in Cockshutt, and I guess there were once more. The old British name for Hereford was *Caerffawydd*, the City of Beech.

JANUARY

Heartwood

Cockshutt, a wood saved by poverty – a modest English wood – transplanting ash – 'of alders' – ageing – the elder, and its evil reputation – a storm – robin song – 'a real wood. With blood' – the foxes' earth – dog's mercury, indicator of ancient woodland

1 JANUARY: In winter no one wants the desolate woods, so I wander the public footpath through the wood above Orcop church alone. On the lane below, small groups in bright clothes walk in the jaunty way of weekending people having self-conscious fun. Laughter comes up to me.

The local woods are moulded around streams, or perched on hilltops, the places inaccessible to the plough and the hay mower. Old Hall Wood is too steep; Cockshutt, across the valley, is too wet; a wood saved from the blade due to poverty of drainage.

The land spills about; hilly though not actively hostile.

West Herefordshire is a separate country. Hard against the mountains of Wales, fronted by the rich River Wye, a place apart, a compromise between man and nature.

Robin song pinballs around the trees.

From up here I regard my wood, Cockshutt, in perspective. Bigger than a copse. But smaller than a domineering forest. A moderate English wood, I decide. The best sort of wood.

2 JANUARY: Frost stars in blue sky; I carry bales of straw to the pigs to warm them. Venus and Mars very clear. Redwings, with their strawberry flank-smear, chattering in the hazels behind the pool.

By the time I finish bedding down the pigs, the moon is in the trees; tonight the branches defend the wood, try to preserve its secrets.

I commit an unintentional pun to Lavender, the lop-eared Welsh pig – 'You can't see a sausage, can you?' She snorts in exasperation.

Old Brown starts hooting; it is about 4pm. He likes to sit in the elegant chestnut tree just inside the wood. (The Romans introduced sweet chestnut. What did the Romans ever do for us? They introduced sweet chestnut.) The tawny owl is, indeed, tawny – a definite brown, but patterned and mottled to resemble the trunks and boughs of its woodland domain. Thus the old country people called the tawny the 'wood owl' or 'beech owl'. Tonight, Old Brown passes himself off as a broken stubby bough.

Old Brown freefalls, only opening his wings at the last nanosecond to stall the impact. I do not see his catch; it is too dim, and he is quick to gobble.

At the top of the wood, now: if I stand on the wood pile I can see the red lights of the TV mast at Checkley; my spiritual centre of gravity, the pivot around which I spent my early life. The place I never leave, no matter where I am.

At night the wood's trees seem to reach forward; shadows break and re-form, dart, become still. Cottage lights on Garway hang in the air.

Gazing at the stars through the wire-work of branches: I decide I must learn the stars, their language; their Morse, their code. All I can manage at present is to join the dots.

3 JANUARY: Mid-afternoon. The silence of a wood, like that of a church, is contagious.

The sheep are lying, heads between front legs, in the way of dogs. Usually, they baa. Not today. I take their cue, and do not open my mouth to call to them. A robin solos, but his voice only stresses the sacred quiet.

I've confined ten Hebridean sheep behind wire fencing tied to trees with the inevitable baler twine. Hebrideans, small, black and primitive, will eat, thrive even, on the iron rations of bramble leaves, much as wintering deer do. They are bramble-destroyers with horns on. Where bramble grows hardly anything else grows.

I am making a last glade, of about a quarter of an acre near the top of Cockshutt, thin into the oaks like a sword. There was no variety in the wood when we arrived; it was an unlimited sea of briars crashing around trees, except for blank lands under the beech and conifers.

5 JANUARY: Wood work. Transplanting ash 'whips' and saplings from the bank of the dingle to the grassy banks below the old ride. Sometimes even the fittest trees need human help to survive; the saplings were too dense for their own good.

Ash has lost its importance to mankind. In the early nineteenth century one British farmer declared, 'We could not well have a wagon, a cart, a coach, a wheelbarrow, a plough, a harrow, a spade, an axe or a hammer if we had no ash. It gives us poles for hops; hurdle gates wherewith to pen in our sheep; and hoops for our washing tubs.'

John Evelyn adduced more uses of ash in his *Sylva*:

It serves . . . the scholar, who made use of the inner bark to write on, before the invention of paper, &c. The carpenter, wheel-wright, cart-wright, for ploughs, axle-trees, wheel-rings, harrows, bulls, oares, the best blocks for pullies and sheffs, as seamen name them; for drying herrings, no wood like it, and the bark for the tanning of nets; and, like the elm, for the same property (of not being so apt to split and scale) excellent for tenons and mortaises. Also for the cooper, turner, and thatcher: nothing like it for our garden palisade-hedges, hop-yards, poles, and spars, handles, stocks for tools, spade-trees, &c. In sum, the husbandman cannot be without the ash for his carts, ladders, and other tackling, from the pike to the plow, spear, and bow; for of ash were they formerly made, and therefore

reckon'd amongst those woods, which after long tension,
has a natural spring, and recovers its position; so as in
peace and war it is a wood in highest request.

This 'useful and profitable tree' also gave the 'sweetest
of our forest-fuelling, and the fittest for ladies chambers',
while the leaves were relief to cattle in winter.

The Ancient Egyptians imported ash from Europe
as far back as 2000 BC to make wheels.

I'm spreading the ash throughout the wood in the
hope that some will beat the odds against ash dieback,
caused by the fungus *Chalara fraxinea*. That Britain's
80 million ash trees will not suffer the fate of the elm,
whose demise transformed the landscapes beloved of
John Constable.

The French soldier Marshal Hubert Lyautey (1854–1934)
once asked his gardener to plant a tree. The gardener
objected, saying trees were slow growers and it would
take a hundred years for the tree to mature. The old
soldier replied: 'In that case, there is no time to lose;
plant it this afternoon.'

Of alders: there are four mature alders in the wood, and
six in the pig paddock, all twenty feet in circumference.
As the 1903 Ordnance Survey confirms, the paddock

was originally part of the wood, but only the alders still stand. All of them are coppiced into hydra-headed giants, of eight, nine, ten uprights each.

The alders honour-guard the peaty streamlet. Alders love the secret, mirey places. To slush through the black, oozing earth between the paddock alders is to enter a druid's grove.

Like the ash, the alder is yesterday's tree. Evelyn positively rhapsodized them:

And as then, so now, are over-grown alders frequently sought after, for such buildings as lie continually under water, where it will harden like a very stone . . . they us'd it under that famous Bridge at Venice, the Rialto, which passes over the Gran-Canal, bearing a vast weight. Jos. Bauhimus pretends, that in tract of time, it turns to stone; which perhaps it may seem to be (as well as other aquatick) where it meets with some lapidescant quality in the earth and water.

The poles of alder are as useful as those of willows; but the coals far exceed them, especially for gun-powder: The wood is likewise useful for piles, pumps, hop-poles, water-pipes, troughs, sluces, small trays, and trenchers, wooden-heels; the bark is precious to dyers, and some tanners, and leather-dressers make use of it; and with it, and the fruits (instead of galls) they compose an ink. The fresh leaves alone applied to the naked soal of the foot,

infinitely refresh the surbated traveller. The bark macer-
ated in water, with a little rust of iron, makes a black dye,
which may also be us'd for ink.

The dye from the alder flowers stained Robin Hood's
clouts green.

Cockshutt's alders, which are about two hundred
years old, were perhaps grown for the local building
industry. More probably, they were a source of firewood
for the lime kilns on the lane to Bagwyllidiart (an un-
pronounceable Welsh name left behind from when this
was disputed land: it is simply 'Baggy' if you are local.
To the amusement of my friends, the first person they
ever encountered who could say 'Bag-ill-id-ee-art' was
my wife. Who is from London.)

It was the Romans who developed the burning
of limestone to make lime as the main ingredient in
mortars, concretes, plasters, renders and washes. During
the Middle Ages the demand for lime increased as the
construction of castles, city walls and religious buildings went up. House walls were lime-washed to make
them waterproof and as decoration to brighten and dis-
infect the interiors.

Lime kilns were used increasingly for the production
of lime for agricultural purposes. When added to acidic
soils, like the Devonian red sandstone of Herefordshire,
lime broke up the clay, 'sweetened' the grass, suppressed
weeds, prevented foot rot in livestock.

Ah, lime: the cure-all, in a long list of putative cure-alls sold to farmers. The advertising slogan for agricultural lime was 'Give Life to the Land'. Farmers eventually learned the painful adage, 'Lime enriches the father but impoverishes the son.' Lime's improvement of the soil was transitory, meaning ever more was needed to keep up productivity.

Farmers with larger landholdings had their own lime kilns, and burned limestone for others. At a price.

Cockshutt and the other local woods produced the fuel (possibly turned into charcoal first); the limestone, or chalk, came from Woolhope ten miles away.

For every two tons of limestone burned, a ton of lime was produced. Lime kilns entered their demise at the end of the nineteenth century, when chemical fertilizers began their resistible rise.

The kiln on the lane, which is built into the bank, is a flare kiln, with rough, squat stone chambers, formed into hollow eyes. Today, the chambers are full of 7UP cans.

How do I know the age of the alders? One does not need to cut down a tree and count the annual growth rings to know its age. This can be done non-invasively.

Begin by determining the tree species, then measure the circumference in feet using a tape measure at diameter breast height ('DBH' in the jargon) above stump level. Then divide the circumference by 3.14 (pi) to find the diameter. Multiply the tree's diameter by its growth factor as determined by species (see the list below) to find the tree's approximate age.

Alder – 5.0 Growth Factor × diameter
Birch – 5.0 Growth Factor × diameter
Ash – 4.0 Growth Factor × diameter
Cherry – 5.0 Growth Factor × diameter
Oak S – 5.0 Growth Factor × diameter
Elm – 4.0 Growth Factor × diameter

Easier yet, the late dendrologist Alan Mitchell found that most trees conform to a simple rule: mean growth in girth in trees in a wood, where there is lean competition for light, space and food, is an inch a year. Trees in parks and gardens grow faster and fatter, and sprawl like sultans eating Turkish delight on divans. So a tree in open ground with an eight-foot girth is about a hundred years old – a tree of similar size in a wood might be two hundred years old. There are exceptions to Mitchell's general rule of tree ageing; most young trees grow much faster – growth slows with age – so in very old trees the growth rate over the life of the tree will be less. In certain species, such as Cockshutt's giant redwood, normal growth is two to three inches per year.

Measuring trees is addictive. The first prolific tree measurer was John Claudius Loudon, whose impressive eight-volume *Arboretum et Fruticetum Britannicum* (1834–37) provides us with more than five hundred historical tree measurements. Between 1880 and 1895 Robert Hutchison measured nearly a thousand trees, mostly in Scotland, and more than 3,500 records appear throughout the volumes of *Trees of Britain and Ireland* by Elwes and Henry (1900–13). Modern-day tree measuring reached its zenith with Alan Mitchell, who measured more than a hundred thousand trees between 1953 and 1995 and co-founded *The Tree Register of the British Isles*.

Cockshutt's trees were harvested for industry before Georgian times. During the Roman occupation the Wormelow Hundred, of which Cockshutt was part, supplied vast quantities of timber for the purpose of smelting iron ore brought by the Romans from the Forest of Dean. There were forges at St Weonards, just over the hill. (If you live in hill country everything is over the hill.)

7 JANUARY: Mild; winter gnats pester the face, while I eradicate sycamore saplings, thus bringing a welcome sense of space to the bank of the dingle.

Then it starts to rain. My tools are an axe, an ordinary hook, a slasher, a bow saw and pruners. I select them in the manner of a surgeon selecting an instrument in theatre; I too, like him/her, wear rubber gloves – but mine are yellow Marigolds for washing up. To stop the tools slipping.

Every time I cut a tree, drops of rain fall from the branches. Like tears.

Trees: you never really know what they are thinking. John Stewart Collis, 1940s naturalist, farmer, proto-conservationist, observed: 'Truly trees are Beings. We feel that to be so. Hence their silence, their indifference to us is almost exasperating. We would speak to them, we would ask their message; for they seem to hold some weighty truth, some special secret – and though sometimes we receive their blessing, they do not answer. '

I am surrounded by a gang of trees, suddenly malevolent in the rain and the noir sky. It is time to leave.

> Then spake I to the tree,
> Were ye your own desire
> What is it ye would be?
> Answered the tree to me,
> I am my own desire;
> I am what I would be.
> Isaac Rosenberg

8 JANUARY: With the exception of the coppiced Wishbone Oak, the oaks in Cockshutt are uniformly straight and tall, and none has a body thicker than a man. Only the Wishbone is a giant.

A slow sky, lichen grey, the same colour as the oaks this afternoon.

It is getting on for evening when I leave. Deep up in the last oak a female tawny owl *kerwick*s – a casual scraping noise, akin to the opening of a metal garden gate. This is her contact call, and it is becoming the aural grammar of night in Cockshutt Wood, a constant comma. Old Brown has a new mate.

The woodcock are back. The wood breathes birds in, and out.

Woodcock snooze during the day, and come out to feed in the evening, probing the damp dingle floor with their long beaks. They can detect worms and other underground creatures by the sensitive nerves at the end of the beak. Sometimes they will creep out when it is still light. As they did today. I saw them.

10 JANUARY: Very mild; but a black voice of crow at the north of the wood. The Jew's ears have 'inflated' – four hundred of them, in size from buttons to porcine ears, on the elder by the stile.

With its fissured trunk and tendency to loll, the elder tree is not a thing of beauty. But from such arboreal ugliness emerges in early summer a generous white flower that is glory to eye and nose.

The elder thrives almost everywhere, though it likes nitrogen-loaded soil best. A sort of proof: the stile elder sits on the same pink clay mound (the remnant root system of a holly upturned an aeon ago) the rabbits use as a sentry post, and which is encrusted with their droppings.

Judas tree. Devil's wood. God's stinking tree. Black elder. Bour. Scaw. Dog tree. The cruel country names for the elder reflect the belief that Judas hanged himself from the elder. It is the devil's own wood, and if burned in Warwickshire you'll see Lucifer himself pouring down the chimney. The Scots, among others, considered that the Cross of Calvary was 'bour', making it, dreadfully, symbol of both death and sorrow:

> Bour-tree, bour-tree, crookit rung,
> Never straight, and never strong,
> Ever bush, and never tree
> Since our Lord was nailed t'ye.

Some of the evil reputation attached to the elder arose from its habit of killing the peasants of the Dark Ages as they shivered in their rude hovels. Elder wood releases cyanide when burning.

Sambucus nigra has hard wood but hollow stems, the easy source, in times now long gone, of flutes, pipes and pop-guns. Etymologically, elder is maybe stemmed from *Æld*, 'fire' in Old English, due to the employment of its branches when blowing on nascent flames. The Latin name, at the experts' guess, comes from a musical instrument fashioned from elder wood, called a sambuca.

Poor elder. Its youth is green and too brief; there is no adolescence, no pert twenties, MILF/DILF thirties – it goes straight to wrinkly, twisted-bone, deep old age, and loiters about, unloved, for decades, losing its bark in an arboreal psoriasis.

The elder is a miserable tree. Collis thought it 'hopelessly plebeian', a bush posing as a tree, a tree failing to be a bush.

13 JANUARY: Snow last night, which then froze.

I wake up to a white paradise.

Snow remits the usual laws of life; a snowy day is a holiday.

At the wood: vertical stripes of snow on north-facing trees. And deer prints on the horizontal path. A wood exists in plural dimensions.

I follow the slots in the snow, lured on by the hope of spotting the deer, but the marks trot out of Cockshutt over the neighbour's wheatfield and away into the polar distance, probably to Hole Wood. I have heard fallow deer barking in Hole, a thirty-acre black duvet tucked down in the valley.

Back in Cockshutt. There are ten blackbirds in a flock, the largest congregation of the species I have seen. They squawk around the wood, fiercely, like outraged ayatollahs.

The snow hardly dents the ardour of the animals in heat: a grey squirrel couple hump unashamed against the Tall Oak. Yesterday, I caught the fox pair entwined. Dogging, you might say.

The fox, Britain's only wild canine, has inhabited the isles for tens of thousands of years. It became native at the time of the Ice Age, originally sharing its domain with woolly mammoths and sabre-tooth cats. So, I was the interloper in the scene.

14 JANUARY: Mid-afternoon. The piglets are at the stream in the dingle. I think of them as domestic animals; they surely think of themselves as wild. A matter of perception.

I hold the face of one of the Large Black sows, raise her ears, and look into her brown eyes. You can perceive the intelligence. Pigs have locked-in syndrome. They have the thoughts; they cannot express them verbally.

A raven flies overhead, unseen but marked by the distinct sound of its wings; a horsewhip flicking, a stick swishing.

15 JANUARY: A storm in the morning; and I'm in the centre of Cockshutt, which is moving like the deck of a schooner going round the Cape.

Timbers splinter; surf pounds through the crows' nests; bines of honeysuckle shriek.

Force 10. Wind so oxygenating I can hardly breathe.

We hold on, me and the exhilarated trees.

But the beating of the wind and the rain drives the birds out of the wood, blasts away the blackbirds. One somersaults over and again, a black handkerchief in a tumble dryer.

By afternoon the wind is pining for its former fortitude.

16 JANUARY: Heavy with meltwater, the stream through the dingle and the paddock is running faster than a man can walk and at the bottom the overspill has formed a small lake. The alder emerge from the water like the masts of sunken Sargasso wrecks.

Curious: the old maps show that this new lake is where the wood's pool was located before it was drained, and relocated deeper in Cockshutt by damming the dingle gorge with rough-cast concrete blocks in the 1950s.

Meet the new pool, same as the original pool.

The wind ploughs long furrows in the flood. The same wind carries the bleats of trapped sheep, and the pitiful *peeow*s of circling buzzards over in the next valley.

In the morning's flat sunshine, the flood is as silver as sea. Although the flood-pool has only existed for hours, there are three mallard sailing there, tacking into the wind. A fine picture.

The birds spark a sudden memory: the Tuileries Gardens, Paris, our children playing with wooden yachts on the pond.

Behind the ad hoc pool, a kestrel quarters the winter kale, slipping forward in small curves, then anchoring to fan the ground into stupefied submission.

There are long-tailed tits working the hazel trees of the hedge, tinkling as they go, minute feathered wind-chimes.

I get to within forty feet of the duck before they launch, alarmed, up into the wind, a near-vertical take-off.

Left behind on the water: a bobbing, bountiful tide of hazelnuts. Where did they come from? Are they a jay's treasure hoard, stolen away by the flood? Distinctive spoors of a grey squirrel in the mud, two big feet followed by two small feet in potato-print regularity, proves it has helped itself to this unexpected gift. Grey squirrels do not properly hibernate. They are up in a drey I have not located, slothsome resting.

The sky turns sullen. But under the hedge, a single snowdrop has pierced through the earth and bloomed. If snowdrops are appearing, then the earth must be

71

wakening. Of all our wild flowers the white bells are the purest, the most ethereal, the most chaste.

How the snowdrop first came to Britain is impossible to know; it was not recognized as a wild plant until the eighteenth century, and it is only here in the west that it flourishes naturally in the woods and hedges.

Whatever; the snowdrop says that winter is not for ever.

17 JANUARY: 9am. The wood is full of sad robin soliloquies, and rain. The notes of birdsong drip down with the drops of water.

Globules of rain stick on the end of twigs; fairy lights.

Midday: I go into the wood deliberately, to kill and to forage, to see what in the wood might sustain stomach as well as soul. I shoot a siesta-ing wood pigeon in the larch, with the .410 shotgun. The smell of cordite hangs in the air for an age.

I farm for wildlife. Cannot wildlife provide me with a meal? Is that not fair? Or at least a decent bargain with nature?

So, this is a real wood. With blood.

Wood pigeon. Due to its generality, one forgets it is a woodland bird. Unlike a lot of native birds the wood pigeon is a proper herbivore and has a crop like a pheasant. It is a grazing animal, and under natural woodland conditions it feeds on the ground in clearings, and come the autumn has a craving for acorns.

A heron sky-rows past. Like a headmaster, it has the eye for spotting miscreancy. The heron *kranks* the alarm. A chaffinch flings itself into the protected space of a hawthorn bush.

All through the wood, scarlet elf cup mushrooms erupt through the floor. It is a good year for *Sarcoscypha coccinea*, which are passable when quick-fried in oil. In times past they were also made into arrangements for the table.

Slugs like them too.

The rabbits in their winter coats have stopped hunger by gnawing the bark of the larch. They do not venture from their burrows this day. Some caution, in the ether or in the DNA, has warned them about the man with the gun.

There is an aviary of twittering in the firs at the very top of the wood: goldcrests.

Back down the path, I am struck by the self-possession of trees, but also their stationariness; I am never more aware of my mobility than when walking past a tree. I leave them behind.

No, they leave me behind; a tree can live for a thousand or more years. Me? A few decades if I'm lucky, and good. (The Fortingall Yew in the churchyard of the village of Fortingall in Perthshire, Scotland, is upwards of three thousand years old.) I might only make a day more than the four decades already won.

At the end of the day: the low winter light, the trees reduced to their bare bones, and there is only loneliness and the scent of decomposition.

19 JANUARY: The rush of wind in spruce. A primordial noise: the first conifers appeared in the Permian period, an age before the dinosaurs. When the wind is in the spruce you hear the world 300 million years ago.

Conifers flourish in conditions that flowering plants find difficult. Conifers are indicators of bad land; they mark poverty, as sure as ragged clothes, holes in shoes. On good soil, conifers tend to be ousted by angiosperms, the class of trees whose mature seed is surrounded by the ovule. Broadleaved hardwood, in other words. Conifer means 'cone-bearing'. All cones are either male or female; never hermaphrodite.

In nature you never know what you will find. Nothing is certain. Especially in a wood. Some years the beech produce no mast; once the woodcock failed to visit.

Birch twigs were traditionally used for punishing: the generic name *Betula* derives from the Latin for 'beat'. I walk into a branch, half lopped off by the wind, which cuts my face. In their dying, trees have the quality of reshaping. Thus, of surprise.

Hazel catkins (lamb's tails) dangle gently from the

twigs, turning lemon yellow from lime green as the pollen forms in them.

21 JANUARY: Near the end of day, and near freezing. Fog clutches the ground; fog so dense there is no demarcation between pond, wood, sky.

Crouched beside the pool, the water coal-slack black. Waiting. Yesterday I spotted strange, obnoxious bird faeces by the pond.

High in the oak, above the fog level, the robin whistles; the low-living blackbird is suffocated into silence.

I hear incoming geese honking, and my worst fears are confirmed. A pair of Canada geese have discovered Cockshutt's pool.

23 JANUARY: A fine morning, with blue tits doing early house-hunting. Lesser celandine out. A young song thrush practises his lines. In low sunshine through the larch a 'lek' of winter gnats dance, as if moving up and down on unseen elastic strings. These are male *Trichocera annulata* seeking females.

24 JANUARY: Heavy frost overnight, which bites into the earth, and I wood-wander in echoey isolated tones of my own making.

Janus was the two-headed god of vigil. Sitting in my chair I look back on the old year, with its broken hogweed stems, collapsed nettles, and look forward to spring, with its green dog's mercury. And the dog's mercury (*Mercurialis perennis*) has risen to two inches.

Later: I spend an hour exploring the eastern edge, the long ditch, which is overflowing with the stink of vixen. I follow the path to their earth. They are the wombling foxes, making good use of the things that they find; the entrance to their earth is under a dystopian dump of corrugated iron, a woman's bicycle, a child's swing, tractor tyres bulldozed there by some previous farmer years ago.

Their metal-roofed den is on the dry bank, up above the piddle stream which runs in and out of the pool.

Pigeons come home to roost. Forty of them.

27 JANUARY: Night in the winter wood: moonshine splays the trees, creating endless plots of light and shade. I hear the badger coming down the path before I see him. Badgers like to sleep through cold periods, a partial hibernation, although the rule, as now, is not absolute.

The badger has found an easy source of food: leftover wheat concentrate pellets (sow rolls) in the pig troughs.

When I first walked the wood's path and polluted it with my man-scent, the animals avoided it, and tried to make their own parallel course. After three months or so, I became as familiar to them as the earth, the sky, the water, and they reverted to the old way.

28 JANUARY: Still the cold continues.

Wind crinkles the pond into sharp waves, making the geese yo-yo.

I spend the morning coppicing hazel, arm-aching work with a bow saw.

Coppicing is a matter of timing. Cut too early and the stump produces shoots that do not harden before frost; cut too late and it disturbs nesting birds, and the tree will have used precious energy on sap which it needs for regrowth. So January it is.

The hazel in Cockshutt was coppiced for sticks for sheep hurdles, forage and bean sticks for the kitchen gardens.

The to-fro rasp of a saw in woodland seems profane, as though sanctity is being violated; but it is rather the music of consideration, of a wood being cared for.

Hazel, unusually, will self-coppice, throwing up plenty of straight rods from the stump. I have a hazel stick, cut from a hedge at Abbey Dore, fifteen years ago, with a V notch at the top, on which to hang the thumb. It is the stick I lean on to gaze at sheep while they safely graze, and wave at them when they have strayed. It is my giant extra arm when I 'cush', move on, the cows.

The tallest hazel in Britain is just over the hill from Cockshutt, at Kentchurch. It is twenty-six feet tall.

Small, bright leaves of wood sorrel emerging on the dank dingle bank.

Wood sorrel's leaves are perfect green hearts, carved out of jade by some master craftsman. The leaves are hypersensitive and mobile; they close when exposed to bright light, when it rains, and when night falls. (The flimsy flowers do not open until April.)

The heart-shaped leaves of sweet violet are also shoving aside the leaf mould. Two male squirrels chase a female round and round, chattering, leaping.

Coal tits singing their spring song in the larch, a more liquid, soprano take on the great tit's *tis sweet, tis sweet*.

In the woodland ecosystem, dead wood plays a godly role. Rot is good. In Britain, the rarest and most threatened saproxylic invertebrates (those dependent on dead or decaying wood) are found in historic parkland and wood pasture. More than two thousand different invertebrate species in Britain are wholly reliant on corpsewood.

So, the 'brashings' from the coppicing are piled up, a small hill of branches and twigs. There are five other log and brush piles in the wood, composed of the leftovers from previous clearance work. All have been colonized by algae, mosses, fungi, wood-boring insects, toads, and mosquito-like flies called fungus gnats. These are weak flyers, and crawl in their hundreds over the old brush piles. So many blue tits are enjoying these easy pickings that the heaps are bejewelled blue and yellow.

30 JANUARY: For a change, I enter Cockshutt via the field gate and the ride, and immediately stumble over an animal whose identity requires me to scroll through an internal ID chart, because it is still, and does not move. Some sort of pet? A guinea pig?

One is unused to wild animals underfoot; the bundle is a rabbit with myxomatosis. I go and get an air rifle from the Land Rover, and put the mite out of its bleeding-eyes misery.

The rabbit came from the small colony under the briars in the ride; there are five other colonies in the wood, all small, no more than ten adults each.

31 JANUARY: Third day of rain; Jew's ears rehydrated again to enormous degree, thirteen centimetres by nine. The dog's mercury is three inches high.

Dog's mercury is an indicator of ancient woodland, woodland which has existed since AD 1600. Various taxa can be used to give an indication that a site has been continuously wooded for more than four hundred years; these include invertebrates (such as beetles associated with dead and decaying wood) and vascular plants (in conservationist jargon 'AWVPs', ancient woodland vascular plants), the species which tend to occur mostly or entirely in old woodland. One singular reason why AWVPs are more common in ancient woods is that they are glacially slow colonists, with relatively poor dispersal abilities. Of ancient woodland indicator plants Cockshutt contains, in addition to dog's mercury: woodruff, wood speedwell, bluebell, wood anemone, opposite-leaved golden-saxifrage.

I planted the wood sorrel.

FEBRUARY

Roots

The warmest tree to touch – how trees grow – wood
pigeon courtship – my first memory – Edith dies – the
mating song of the male chaffinch – catkins – the
subterranean life of trees – sheep on the ride – the fox –
the shooting of the Canada geese

2 FEBRUARY: 8.15am. According to the thermometer
on the barn wall it is 2°C. I drive the six miles east to
the wood and touch the trees: the ash, the beech, the
oak, the birch, the sallow – the elder is the warmest of
them all, followed by the larch.

On this day in 1787 Gilbert White recorded at
Selborne: 'brown wood-owls . . . sit hooting all night on
my wall-nut trees. Their note is like a fine vox humana
& very tuneable.' A friend of White's found tawnies, all
tawnies, hooted in B-flat.

4 FEBRUARY: On the lane. Looking through the Land
Rover window towards Cockshutt's iron oaks, standing
their ground against the rain. I have been defeated and
left my fencing work, but the oaks are uncompromising.

In the 1990s, I rediscovered (after hazy, lazy univer-
sity days) graft, by taking up the family line, farming.

I can hack days of work outside in rain. I tell you
that. In 2001 I spent six consecutive days, on the hill
above Abbey Dore, in mad ceaseless downpour, building

a 'race' (a corridor to contain cattle). Every hole for the uprights, which were railway sleepers, filled with water before I had finished digging it. I did twenty holes, hauled every sleeper myself because it was so slick the tractor's wheels would not grip. The horizontal rails, motorway crash barriers, I fixed myself.

Today, I have given up, bowed before the elements. I am flesh. The trees are wood. We are not the same.

The Hebridean sheep, also made of flesh and tough, have sheltered behind the boles of the oaks, which are bulwarks.

Where does tree growth come from? How can the trunk increase in thickness and be continuously functional?

Xylem is plant vascular tissue that conveys water and dissolved minerals from the roots to the rest of the tree, and provides support. In trees, xylem builds a ring of new xylem around itself. Dead xylem becomes heartwood; and the newer xylem outside, still serving as plumbing (like the Pompidou Centre), becomes 'sapwood'. The heartwood provides the backbone of the tree, the stick for the lolly. The age of a tree may be determined by counting the number of annual xylem rings at the base of the trunk, when cut in cross section.

8 FEBRUARY: Owls still hunting at 7am.

At 8am, I begin laying the western hedge of Cockshutt, cutting each 'stemble' (as the old-timers would call a stem) to within a quarter of an inch of its life then laying it down at thirty-five degrees. There are four degrees of frost. Morning notes of a blackbird in the ash skip off the white surface of the adjacent wheat-field, as thrown flat pebbles skim over water.

My hands thick with leather gauntlet, I drop the big axe. There is no thud; instead the earth rings with the metallic toll of a bell. I let the rest of the medieval executioner's tools fall to the ground. Small axe, bill-hook, bow saw. The music is atonal, Stockhausen.

By ten I am hot, my face rinsed with sweat, and not just from the hedge-laying. The day has gone warm. The thawing of the earth frees the salad smells of new life.

Over the thicket, a pair of wood pigeons rise and fall in flight, oddly like the swooping paper darts one threw at school. Such is their courtship routine. Country folk used to believe that Valentine's Day was for more than humans; it was the day the birds became betrothed. Chaucer wrote a poem, 'The Parlement of Foules', on the very theme.

The love of birds is very British, very old.

There is an unexpected privilege in hedge-laying; one gets to peep into the inner sanctum of this linear woodland before the harem veil of leaves comes down.

You see the hedge's secret life. Chopping into the base of a hawthorn, properly almost severing it but not quite, the billhook hits a patch of glossed bark, and swings back with two or three ginger hairs stuck to the sappy blade. Whether from weasel or stoat I am not sure, but I have found the passageway of a mustelid.

Hedge-laying is one of those jobs that place one in the river of all human time. I'm cutting at wood with a metal billhook; the first Neolithic farmers here chopped at the wildwood with flint adzes.

Different cutting medium; same perspiring action.

There is another antique element in the scene. Man and dog. I have Edith with me, and she alternately mooches in the hedge and then lies in the sun with matronly dignity.

I have known her since the moment she was born. I watched her enter my world. She has always been the most beautiful black Labrador, outside and in. She would be a shoo-in for a canine *Country Life* 'girl in pearls'. Indeed, her proper name is Edith Swannesha, 'Edith Swan-neck', in honour of the legendary Anglo-Saxon beauty married to King Harold.

My Edith is dying. She has cancer, and I am atomic-clock counting down the days of her life. Tick. Tick.

Tick.

*

My first memory: I am a toddler, strapped in a Silvercross pushchair, the one with the green canopy to keep out the 1960s sun. My parents (I learn later) are on holiday in Venice, and I am in the care of my great-aunt and great-uncle, Kathy and Willi, who farm sheep at Llangennith on the Gower.

The pushchair is on the stone flags outside the front door of their lime-washed (with Snowcem) farmhouse. Shadow, Willi's sheepdog, slinks up to the side of the pushchair, levers himself up and peers at me. Before biting me on the face.

I was bitten by more than a dog that afternoon, I was bitten by a bug: I have loved dogs and farming ever since.

I'll explain. It was, of course, my fault that Shadow bit me; I had provoked him by waving an ice cream with a 99 chocolate flake in his face. I would have bitten me too.

Shadow was my brother – I did not have a human one, or a sister – and on that summer afternoon in the 1960s I learned to treat him like one: with awe. Modern biologists reveal almost monthly that the 'species barrier' between animals and humans is thin; I learned that, Mr Scientist, in my pushchair.

I think it was the same afternoon that I saw Shadow flush sheep from the vague ruins of the castle on one word of command from Willi. I held Willi's wind-tanned hand; transmitted through it was the Zen contentment of a man in harmony with his dog, his stock, his land. I wanted the same. I have always wanted the same.

10 FEBRUARY: For another day the black oaks have stood in squalling rain, straight as the masts of naval ships. They did not bow, they did not yield.

Edith dies, and I howl.

11 FEBRUARY: I listened to the sweet south wind this morning . . .

12 FEBRUARY: Chaffinch singing in the cold winter sunshine, which is then dashed by sleet. He has acquired a blue cap, and a rich pink waistcoat. This is the first full mating song by a Cockshutt bird.

13 FEBRUARY: Sunny, which brings the forgotten delight of sunshine on water, and gnats jigging.

I poke, with a stick, at the decaying ash tree in the dingle, the one which passes for a perpendicular split banana, revealing its rotten, crumbling grey core.

Trees age in their way, not ours. The decomposing heartwood, the fallen twigs, branches and leaves all go to feed the tree. In order to keep growing upwards and outwards, the tree cannibalizes itself. By this means, it can reach an effective immortality, for how else would you categorize a life of a thousand years or more?

Trees first evolved on Earth around 400 million years ago. They survived the meteorite that did for the dinosaurs.

Britain is home to more dead-wood invertebrates than any other European country for the simple reason we have more veteran trees.

The lords and ladies, with their poxy leaves, are four inches high.

A kestrel at the edge of the wood: the humming bird of prey, vibrating on the edge of ecstasy.

Atop the rabbits' sentry stump are hazelnuts split clean down the middle, the forensic proof of the grey squirrel, which gnaws a small hole in the nut, then bisects with bottom incisors to make a clean break along the manufactured weakest line. Small neat holes in nuts are made by wood mice or voles.

The male catkins on the alder have expanded from hard chipolatas to soft lengths of pipe cleaner.

My last pig joke: I ask Lavender if her favourite painter is Francis Bacon.

Actually, I have one more piece of porcine humour. The pigs like to itch against the rough bark of alder, making, I suggest, 'pork scratchings'.

The pigs also chew at the alder. Alder, when bitten or cut, turns a striking red – the colour of blood. The tree was thus deemed human by the native tribes of Britain and revered as the water spirit's sentinel.

15 FEBRUARY: On/off rain. I spend the morning chopping down, and de-limbing, two spruce trees to repurpose as poles for the repair of the corrugated-iron pigsty. White chips fly, and the scent of pine resin fills the morning.

On the larch, the leaf buds sit in fat little pods along the dry brown twigs; soon the fresh green needles will emerge. The honeysuckle is already in new leaf.

Underneath the beech, a young beech rises from the ground, as muscular as a conger eel.

To trim a tree flush with the trunk is 'snedding', same as it was in Anglo-Saxon England.

Under the woodland floor, the unseen life.

Trees are networkers, not solitary acts. Trees communicate with one another through underground fungi – mycorrhizae, which weave into the tips of their roots and extend through the soil to form a subterranean internet, via which can be sent nutrition (sugar, nitrogen, phosphorus), along with warnings about such crucial arboreal matters as aphid attack. The so-called 'Wood Wide Web'.

The relationship between these mycorrhizal fungi and the trees is old, and mutualistic. The fungi siphon off food from the trees, taking some of the carbon-rich sugar that they produce during photosynthesis. The plants, in turn, obtain nutrients the fungi have acquired from the soil, by use of enzymes that the trees do not possess.

It is not a case of every tree for itself. In the Wood Wide Web a young seedling in a heavily shaded understorey might be supported with extra resources by its stronger neighbours.

But I have an axe to grind. Contemporary arborists are tempted to portray the society of trees as a hippie, communistic utopia (hence the 'Wood Wide Web' in which the trees have personhood, like the Ents in J. R. R. Tolkien's *The Lord of the Rings*. Tolkien's Ents held ponderous conversations that ignored paltry human timescales, and were capable of love.

Trees are not Ents, and to suggest they are as sentient as primates is to diminish them, and us.

Every generation looks at the trees, and takes what it wants. Once, we looked at the trees and considered ourselves and them heroic individuals.

'Ent', incidentally, is Old English for 'giant'.

19 FEBRUARY: Move the sheep down to the ride and the far bank of the dingle, confining them with a hundred metres of electric netting. Grass needs cutting in order to thrive; the wood tries ceaselessly to 're-forest' all areas, and untended grass would become scrub. There are days when I seem to be fighting the wood.

On this same day, I herd the four cows out of the wood and the pigs to the adjacent paddock (which itself is partially wooded; pigs have the devil's piggy appetite for bluebells, and if left in Cockshutt would consume every last one). For the first time in months there are no livestock in Cockshutt, and the wood has lost a little of its soul.

20 FEBRUARY: Once when Cockshutt's trees were silent and white under snow a woodcock fell from the sky, black and fast like a comet from heaven. The trailing sparrowhawk, having misjudged the attack, crashed into the tatty Gothic drape of snow-covered honeysuckle.

After a moment or two, the wriggling sparrowhawk fell to the ground beside the woodcock, but, out of its sky element, unable to attack, flew off.

I watched the wood fill with white that day, until the petrified woodcock became a sugar bird.

*

Outside their metal grotto, the foxes have left the skin of a lamb.

22 FEBRUARY: Frost on snow. Each and every tree in the wood is made individual by the paleness of the backdrop.

Each day, a change, a note to make. A female blackbird, balled up against the cold, searches hopelessly through the sycamore leaves, all stiff and curled. She sees a thin, red twig, which she must have realized was a twig but hopped to it anyway and picked at it. Such was her hunger, and her imagination. I have some apples for the pigs, and throw one her way.

A shrew resorts to hunting in broad daylight.

Afternoon: the disconsolate dog fox, everything about him diminished and dejected, tippy-toes into the pig paddock.

I accidentally let go of the bucket. The fox thins, his mask switches left, then right, sees me. He appears to be smiling. He runs, but he will be back.

We both know this.

23 FEBRUARY: Thaw, due to sunshine. The seasons steal days from each other. Today it is summer in winter. Along the ride, my footprints fill slowly with water; the Anglo-Saxon word *rode* meant woodland ride. The display flight of the woodcock, 'roding', is derived from rode, because the birds often seek and follow rides and the edges of clearings.

Woodcock are now amber-listed in the UK.

27 FEBRUARY: Sitting in my chair. Mist. 2°C. A blade of light appears in the east, and the brightness grows over the pool. The fusion of wood and water. Watching the dawn is the never-diminishing privilege of the early riser.

The birds of the wood can wait no longer. Their desire to mate is stronger than the gravity of returned winter. A great tit raises his voice in confirmation.

28 FEBRUARY: The Canadas have attacked everything: the frogs, the moorhens, any mallard which land. I shout at them, fire off guns, get the dogs to bark at them. They will not budge, they will not move. In the desperate end it is the pool's life or the Canadas'.

My son and I decide to shoot them, with heavy BB cartridges, on this bone-bleak day when the birches on the bank give an oddly apposite North American aspect to the scene.

The Canadas are sitting on the water. A 12-bore broadside to the gander . . . I expect the goose to fly, she refuses.

Tris, as startled as I: 'Do I shoot the other one?'

Is it love? Loyalty? I do not know, only that the goose bravely faces the gun. I say 'Yes,' but I am no longer thinking about pest control, only stopping a bird's anguish for the love of its life.

Dead on the water, the Canadas float on their backs, and when the wind blows them towards us they are as menacing as pillows. Hauled on to the bank, the Canadas are such beautiful corpses, their black snakey necks damn good enough for a stempost on a Viking longship; the fine sawtooths of the bills delicate engineering beyond the skill of man.

A Winter Eden

A winter Eden in an alder swamp
Where conies now come out to sun and romp,
As near a paradise as it can be
And not melt snow or start a dormant tree.
It lifts existence on a plane of snow
One level higher than the earth below,
One level nearer heaven overhead
And last year's berries shining scarlet red.
It lifts a gaunt luxuriating beast
Where he can stretch and hold his highest feast
On some wild apple tree's young tender bark,
What well may prove the years' high girdle mark.

The Wood

Pairing in all known paradises ends:
Here loveless birds now flock as winter friends,
Content with bud inspecting. They presume
To say which buds are leaf and which are bloom.
A feather hammer gives a double knock.
This Eden day is done at two o'clock.
An hour of winter day might seem too short
To make it worth life's while to wake and sport.

Robert Frost

MARCH

Buds

*Back to black weather – the first celandine and primrose
– the glade I made – a forager's soup – ivy – birch syrup –
the arrival of the chiffchaff – a blackbird makes a nest
in the elm – BB's Brendon Chase – a traveller in a new
land: spring – bramble leaf miners – sycamore*

1 MARCH: Back to black weather (as Amy Winehouse might have written). With no leaves for umbrellas, the rain trickles down the trunks of the trees like sweated sap. Bluebell leaves are long fingers of green stars.

Like all owls, the tawny is no great home-builder. By choice, the tawny will nest in a cavity in a hollow deciduous tree, preferably wrapped with creeper (hence yet another country name for the bird, 'ivy owl'), with the female scratching a desultory scrape in arboreal debris. Which is exactly what the Cockshutt tawnies have done in the dingle.

The rain has not stopped the flowering of the first celandine and the first primrose, with its intense, saliva-inducing lemon beauty.

2 MARCH: The rain has washed and exposed more of the roots of the alders in the paddock, the gall-like structures which grow on alder roots now clearly visible. These bacteria-laden nodules fix nitrogen, and so improve the soil.

One of my farming neighbours has recently cut down all the alders in his hedge.

Along the silent path; a rabbit shoots out of the bramble, a white-tailed thunderbolt. A blackbird *chinks*; then another, so there is a chain of *chink*ing across the trees and hedges and fields to the village gardens on the hill. The success of garden birds, né woodland birds, such as the blackbird is heart-warming. The blackbird is not a high flyer (bird lover's joke) and the dwarf apple trees and shrubs of the villagers suit it well.

One blackbird begins to sing. Is there a better chorister than the blackbird? When Theodore Roosevelt, the President of the US, came to Britain in 1910 he was aghast at the local indifference to the music of the blackbird. I too am enraptured by this darkling bird, the fluting melodies, the pitch perfection. It is not the devil in the detail; it is the divine. Surely.

3 MARCH: In the rains the woodland plants are springing up, especially the dog's mercury, now six inches tall, creating a long low kingdom of spires in sulky corners of the wood.

Change of plan, and I return sheep to the south glade and this time mob-stock, twenty sheep close confined to eat down the sward which is full of reddy sycamore seedlings.

This grassy glade was laid down three years ago, and is, if I can keep the sycamores at bay, full of meadowland wild flowers (cowslips, red clover, yarrow, harebells, ox-eye daisy and dandelion) and grasses galore (meadow fescue, red fescue, cocksfoot, timothy and rough-stalked meadow grass). It is a picture worthy of Sisley or Burne-Jones in late spring.

No grazing by livestock, no glade.

Ideally I would have put the cows on the glade – different types of livestock graze differently, creating varying micro-environments and sward heights, a mosaic of habitats – but it is too wet to support their weight.

On my way out of the wood I disturb a waddling mallard pair having a constitutional.

The sallow wands are turning green: the colour of spring appears early in their skin. Sallow is the Church's substitute for the palms laid on the path of Jesus. Each year, crucifixes made from sallow twigs are distributed as palm on Palm Sunday. Folklore states that Christ was once whipped with a sallow rod, and if this indignity was visited upon children their growth would be stunted.

I coppiced some of the sallow three winters back;

the sallow shoots now are two and a half metres high. In the same area on the east of the wood I coppiced three ash; the new stems thrash, Medusa style.

4 MARCH: Gale. The claw-ends of the spruces are blown off to create horror scenes of decapitated green hands on the woodland floor. A very small pinch of the fresh pine shoot is pleasant to taste; these shoots, eaten constantly, were once considered to cure chest disease.

The wind bends the ash branches back and forth, like oars on a skiff.

5 MARCH: The elder leaves are out, the first deciduous frondescence; they stink of poison gas. In the old time the elder was planted at the back door, to keep evil spirits and other negative influences from entering the home. William Coles, in *The Art of Simpling* (1656), noted that the 'common people' gathered elder leaves 'upon the last day of Aprill, which to disappoint the Charmes of Witches, they had affixed to their Doores and Windowes'.

The aroma exuded by the elder's leaves certainly repelled flies, and bunches of leaves were hung in livestock barns, and attached to horses' harnesses. Elder leaf, like elder bark, is full of cyanogenic glucosides. The leaf also contains the mind-bending neurotoxic alkaloids sambugrine and conicine, which might explain why one meets the king of the elves when sleeping under an elder tree.

I go gathering scarlet elf cups from logs emeralded with moss, then some Jew's ears from the elder. Both mushrooms have small taste; instead they absorb the flavour of other foods. The trick with these mushrooms is to shred them, thus piercing any pockets of water, otherwise they spit when cooked.

Besides its use in the kitchen, Jew's ear has a role to play in the chemist's cupboard. The Chinese have long valued it as a cleanser of the lungs, stomach and intestines, while modern Western medicine reports the fungus as having significant anti-blood-clotting characteristics, and of likely therapeutic use in combating coronary disease. There is also evidence that Jew's ear has antibiotic and antiviral qualities. In the medieval 'doctrine of signatures' (which held, essentially, that if a plant looked like a human organ it would cure an ailment of the organ), gargles of Jew's ears were made for complaints of the ear.

When we injure woods, we injure ourselves.

Thai-style chicken and Jew's ear soup

My recipe.

Every Jewish grandmother knows that chicken soup cures all ills; every Asian grandmother knows that ginger boosts the immune system. This Thai-inspired broth recipe is a double whammy.

1½ tbsp cooking oil
2 in ginger, peeled and sliced into slivers
1 shallot, diced
5 large Jew's ears

1 skinless and boneless chicken breast, cut into thin strips
1 tsp oyster sauce
1½ tsp shoyu
1 tsp cane sugar
2 tbsp water

Heat up the cooking oil in a wok and stir-fry the ginger slivers. Add in the shallot and Jew's ears and stir. Add the chicken strips. Stir-fry the chicken meat until browned, then add in oyster sauce, shoyu and sugar. Stir all ingredients together before adding the water. Stir several times. Serve with hot white rice.

11 MARCH: Ivy swims up the ash in the dingle, swarming wooden eels. (Ivy is not a parasite, but uses tree trunks for support, to gain height.) Under widow's clouds I feed ivy to the sheep on the ride – they gorge on it, so keen they crowd around me.

I know, I know. Who would believe that ivy is fodder for sheep? It is, except for the berries. Welcome to old-style hard-time farming.

I am trying to learn the music of the inscrutable trees, not just the wind through the branches but the notes of the trunks. I put my ear to a bole of ash; after the seashell noise of the cupped ear, I hear the moaning of a cello.

I look down into the dingle, with its cacophony of dead trees. Dead trees move us, with their truncated majesty, their long, slow decay. At the end of the Great War, the composer Edward Elgar entered a period of disillusion. The values of the Edwardian era had been swept away in the deluge, and his wife, his greatest support, was ill. He himself entertained 'intimations of mortality', to borrow a phrase from Wordsworth. Elgar's last major works, including the Cello Concerto, show this. Billy Reed, the violinist and good friend of Elgar's, recorded that a grove of dead trees near Brinkwells, the Elgars' cottage in Sussex, imprinted the composer strongly at this time:

A *favourite short walk from the house up through the woods brought one clean out of the everyday world to a region prosaically called Flexham Park, which might have been the Wolf's Glen in* Der Freischütz. *The strangeness of the place was created by a group of dead trees which had very gnarled and twisted branches stretching out in an eerie manner as if beckoning one to come nearer. To walk up there in the evening when it was just getting dark was to get 'the creeps' . . . the air of sadness in the quartet, like the wind sighing in those dead trees – I can see it all whenever I play any of these works, or hear them played. Elgar was such a nature-lover and had such an impressionable mind that he could not fail to be influenced by such surroundings.*

Lady Elgar referred to these late chamber works as 'wood magic', and she made direct reference to the dead trees as having an influence on the Quintet. The trees had been struck by lightning (an exterior force) and not died naturally. Just like the young men of the Great War. Elgar made music of England's landscape. 'This is what I hear all day,' he wrote to his friend A. J. Jaager. 'The trees are singing my music – or have I sung theirs?'

My current favourite word: psithurism, *n.*, meaning the sound of rustling leaves on trees, adapted from Ancient Greek ψίθυρος (*psithuros*, 'whispering', 'slanderous').

Go shooting at dusk, while the blackbirds *spink*; a wood is half wild, half tame. To connect with the wild side I take the little .410 and become a hunter.

The fierce-eyed pigeons avoid me. I walk through the larch, and in the half-light it is walking on air; I can't see the floor. Not like at night when one *feels* the floor. In this sort of light one floats.

12 MARCH: Glades of sunlight; the silk-rip of a raven overhead. Primroses leaking spots of sun out of the earth.

A buzzard in the top of a larch mews pathetically. I scuff at some old leaf matter at the top of the wood, and expose corpses of shotgun cartridges. As recently as the 1970s, to judge by the cartridges, Cockshutt was part of a shoot.

I walk back past the pool; frogs are croaking, and the water tremors in ecstasy as they consort, hidden in reeds which now sword the water.

13 MARCH: Hawthorn leaves out, the buds big enough to eat.

I spend an hour up an ash tree cutting ivy for the sheep with a sickle; I feel like Getafix in tales of Asterix and Obelix. But I can smell spring.

Another sure sign of spring: the sheep refuse to eat hay, wanting juicy green stuff instead.

Wreaths from ivy vines were used to decorate the heads of traitors. Such was the fate of the last Welsh prince, Llywelyn ap Gruffydd, whose severed head was presented to King Edward I on a salver, mockingly crowned with a circlet of ivy. In later years a similar wreath, worn by the living, was said to prevent baldness. Now there is a thought.

14 MARCH: A treecreeper walks along the underside of a branch above my head, as upside-down sticky as a fly on a ceiling. In a wood one's eyes naturally focus on thirty to forty yards forward on the ground; but one needs to look up. One has to train oneself to do it.

As I wander through the wood the pigeons are always a move ahead, clattering out of the trees, resiled by the invisible bow waves of my being.

Despite my earlier certitude that trees are not Ents, I look for faces in the oak bark; the only ones I find are the visages of scary old men. The wind gets up. Things crash, fall apart.

On the coffee-water pool, a sudden sequence of disappearing rings. Made by what?

15 MARCH: On the silt around the pond: the triangle toes of pheasant feet, broad-arrow prink of a moorhen's foot, and the half-ring of badger toes.

The badger lives a half-mile away in Moors Wood; this is only the second sign of him rambling in Cockshutt this year. Badger, brock or bawson, call him what you will, he makes me nervous. I have cattle. Badgers are vectors for bovine TB.

Birch sap can be 'tapped' (collected), and used to make a sweet wine.

To tap a silver birch in spring, when the sap begins to rise (as signalled by the development of leaf buds; they should be tight and small), first find a birch tree with a diameter of at least twenty-five centimetres; any less than that, and the tree might be too puny to spare its lifeblood. Drill a hole in the trunk a metre up and at a thirty-degree angle, just penetrating below the bark and no bigger than the plastic tubing you need to stick in the tree to siphon off the sap; the other end of the tubing goes into a demi-john or large mineral-water bottle at ground level. Sap should start collecting in the receptacle immediately; a gallon in twenty-four hours is not uncommon, and take no more than one gallon per tree.

As well as a country wine, birch sap can be drunk as a spring tonic, or boiled down to make syrup.

Writing in 1718 Ned Ward, author of *The London Spy*, described birch wine as 'almost like mead, and makes a man's mouth smell of honey'.

Sycamore sap also rises vigorously in the spring and can be tapped, as a source of sugar and to make beer.

17 MARCH: The chiffchaff arrives in his familiar wood, a day later than last year. True, his piercing, two-tone *chiff-chaff* song is hardly pretty. But it is immortal, and important; it announces the coming-in of spring. As Sir Edward Grey, master ornithologist (and

the Liberal foreign secretary who took us into the Great War), remarked in his *The Charm of Birds*, the chiff-chaff is 'the forerunner of the rush of songbirds that is on its way to us and will arrive in April, and there-after enrich our woods, meadows, and gardens with still further variety and quality of song. That is why the first hearing of a chiffchaff moves us so each spring. He is a symbol, a promise, an assurance of what is to come.'

The chiffchaff hops around a willow, moth-like. How could something so small, such a scrap of bird, manage to fly here? It can barely battle a breeze. Is there not something miraculous in a tiny bird travelling all the way from Africa to Britain to spend spring and summer with us?

How pleased the trees must be to hear him.

18 MARCH: A nuthatch pipes in the oaks, and below, the first wood anemone, white and resplendent.

Legend waxes lyrical about the anemone as a love flower. Mortally wounded by a boar, Adonis lay in the bloodstained grass where he was found by Venus; overcome with grief she swore her lover should live for ever as a flower, and anemones sprang up where her tears fell.

The myth is rather ruined by the anemone's pong; the plant contains proto-anemonin and is poisonous.

But the anemone is beautiful. Traditionally, the first

wood anemones of the year were picked and sewn into one's coat; it was assumed that a flower as exquisite, as pure, as the anemone would, by its sheer goodness, ward off the plague. Some country folk called the anemone 'Candlemas cap', Candlemas being 2 February, the Purification of the Blessed Virgin.

A blackbird begins making a nest in a low fork of elm beside the ride. The elm, which has a hawthorn girding it, reaches into the sky. Even at a mere thirty feet it is the tallest elm I know of.

Elms were one of our defining English trees, regular fixtures in the hedgerow, because the farmer valued them for their shade. Elms grew so rapidly in the west of England they were known as the 'Wiltshire weed' but within a decade, between the 1970s and 1980s, almost all English elms above the size of a shrub were eliminated by Dutch elm disease (caused by a fungus of the genus *Ophiostoma* and carried by various bark beetles), one of the most dramatic extinctions of modern times. Poor elms: for the main part they are clones, and so suffer all the vulnerability of genetic incest.

Elms die when they are teenagers, but are replaced by suckers, the classic elm method of regeneration. Elm had a good reputation as coffin wood.

My elm is merely a small echo of the elms that John Clare knew.

The Shepherd's Tree

Huge elm, with rifted trunk all notched and scarred,
Like to a warrior's destiny! I love
To stretch me often on thy shadowed sward,
And hear the laugh of summer leaves above;
Or on thy buttressed roots to sit, and lean
In careless attitude, and there reflect
On times and deeds and darings that have been –
Old castaways, now swallowed in neglect, –
While thou art towering in thy strength of heart,
Stirring the soul to vain imaginings
In which life's sordid being hath no part.
The wind of that eternal ditty sings,
Humming of future things, that burn the mind
To leave some fragment of itself behind.

John Clare

19 MARCH: Through the druid's grove of alder. Over the stile, struggling with the sack of cattle cake on my left shoulder. Although they are derived from woodland animals, it is estimated that in a totally forested area one square mile of browsing might support only twenty to thirty head of cattle. By simple division the two-plus acres of Cockshutt cannot wholly support four beeves. Ipso facto they require supplementary feeding, hay and a handful of wheat concentrate or 'cake'. Farming: it's all about maths and margins.

The sack weighs me down, so I walk lurched over. A farmboy Quasimodo. From the far top of the wood comes the trumpeting of a cow. Repeated, and insistent.

I could have driven round to the cows as I did this morning, but who does not want the excuse to walk through a wood on a mid-March evening? When the unbearable heaviness of winter has lifted. When the days are lighter, longer.

There is an old farming adage: 'March comes in like a lion, goes out like a lamb.' It is a truth my grandfather taught me. Today, out of the wind, in the shelter of the trees . . . you can actually feel a new, baby tenderness in the air.

Spring is here, and there is a spring in my step as I follow the pale path through the trees.

For me, for you, every step along this woodland path is yard-stoned by the mild English culture of ar-boreality: Shakespeare's *A Midsummer Night's Dream*; *Winnie-the-Pooh*; *The Animals of Farthing Wood*; *The Chronicles of Narnia*; and, of course, *Brendon Chase* by BB (Denys Watkins-Pitchford).

The Germans have Black Forests, and Grimm tales. They can keep them. The folk tales collected and written down in the nineteenth century by the brothers Jacob and Wilhelm Grimm, in the spirit of high Romanticism, codified the forest as pure nature in contrast to the urbane, urban civilization of Germany. In 1876 in Bayreuth Richard Wagner premiered his sixteen-hour operatic tetralogy *Der Ring des Nibelungen*. Inevitably, most scenes took place in a forest. Inevitably, the Nazis were bewitched by the *Hochwald*, the high forest, with its ranks upon ranks of standing, regular pine trees. The Nuremburg rallies were the forest made flesh. Hitler's deputy, Hermann Goering, no less, declared from his forest hunting lodge: 'We have become used to seeing the German nation as eternal. There is no better symbol for us than the forest, which has been and always will be eternal.'

The Anglo-Saxons, on fleeing Germany for England, merely continued the local tradition of cutting down the wildwood to let in the light.

The Other (excerpt)

The forest ended. Glad I was
To feel the light, and hear the hum
Of bees, and smell the drying grass
And the sweet mint, because I had come
To an end of forest, and because
Here was both road and inn, the sum
Of what's not forest . . .

Edward Thomas

I came late to the love of woodland, even to a love of books about woods. I think I was twelve when I read

Brendon Chase, BB's adventure story about the three Hensman boys, Robin (fifteen), John (thirteen), Harold (twelve), who flee the 'petticoat government' of Aunt Ellen to live feral in a wood.

It was *Brendon Chase* that first introduced me to life in a wood, the possibility of adventure in its secluded spaces. I'm still that boy. I cannot walk through a wood without a sense of wonder.

And there is a relaxing privacy to a wood. As BB understood, a wood is a desert island – but of solitude, and plumb in the middle of rural England. A wood is an escape.

I have digressed in thought, but not work. I'm half-way through the wood. That cow is still trumpeting, a siren call, leading me on.

The woodland floor is illuminated by flowers. The black mud trapped in the alders' roots is lit by marsh marigolds, the blooms as bright yellow as any sun a child will draw. (The marsh marigold: a true native, it grew on English soil before the Ice Age.)

Everywhere, the delicate wood anemones are grouped in hamlets of white lights.

The sycamore is in leaf; also the elder. And it is now a month since the first *Arum maculatum*, lords and ladies, pierced the woodland floor in bottle-green shards to begin the killing of winter.

Although birdsong has not reached its spring crescendo, there is a steady anticipatory hum, such as you get in an auditorium as it fills with people, and the orchestra tunes up. The greater spotted woodpecker, dressed in his pied and red avant-garde clothing, is drumming out his love song on the skeletal remains of an elm. His rapid rattle, as though his beak has been wound up with an elastic band and let go, is an invitation to females to come breed with him. It is also a warning to other males to keep off his manor. He can be heard a mile away.

From the high ash the cock song thrush pours in his fluty, reflective melodies. How correct Robert Browning was to write in 'Home Thoughts':

> *That's the wise thrush; he sings each song twice over,*
> *Lest you should think he never could recapture*
> *The first fine careless rapture!*

Tennyson described the voice of the song thrush, mimicking the call of the bird:

> *'Summer is coming, summer is coming.*
> *I know it, I know it, I know it.*
> *Light again, leaf again, life again, love again,'*
> *Yes, my wild little Poet.*

There is insect music too. Bees are about, huzzing. The wood is in motion.

At the pond a peacock butterfly flitters over the water, casting a precise facsimile of itself.

I am a traveller in a new land. Spring.

Through the stand of Norwegian spruce . . . Yes, around every bend of a wood is the possibility of surprise. As I pass the Wishbone Oak, I glance right, and there on the parallel path under the hedge is the dog fox, with a rabbit in his trap-jaw. He is on his way home to the wife and demanding blue-eyed cubs.

The fox looks at me, I look at him. We both have burdens; we give each other a knowing, matey eyes-raised-to-heaven sigh.

For a full minute we carry on our perfect symmetry, jogging our parallel paths, separated by twenty yards of bramble, until the fox disappears down behind the hollies.

By now I have reached the glade at the top of the wood, where a heifer is still mooing her big announcement to the world. This morning she gave birth to her first calf, a conker-shiny girl, now arched under her belly and suckling away. The calf looks happier than the cat that got the cream. She actually appears to be smiling.

I certainly am. Spring flowers are blossoming, I am down to wearing one coat rather than two. And a calf has been born.

21 MARCH: Morning, the leaves still countable, and the enormous silence threaded with song.

There are rabbits running every which way. More

ivy for the sheep. From a beech limb, I hack some 'chips' for smoking pork.

22 MARCH: Another day of grateful warmth.

The wood pigeon is surprisingly virile in flight for a dove, and easily confused with a sparrowhawk. In flight, the pigeon loses dumpiness.

The pigeons are nesting in the larch. (Climb up later; peep at their china-white eggs, two as per usual. The white of the eggs is precisely the brilliant white of the wood pigeon's necklace and wing bars.)

About 6.30pm: the light drains away, the heat too. Cold on face and hands. The pheasants late to bed; the stop-ups. Blackbird vespers from a poolside ash. There is a crashing in the bramble; thudding, hitting – and in me, a primitive frisson. A rabbit bolts.

I am here because of my work, checking sheep 'folded' (fenced) to clear brambles.

Overwhelmed by multitudinous sensations. The leaves are coming; from the encompassing fields the child-cries of lambs on thin April grass.

A treecreeper: up it hops, defying gravity.

I lie in the bend of the Wishbone Oak and absorb the brown earth, the breeze in branches, the song of the birds.

Turning my head towards the pond: the utter romance of moonlight on water seen through trees.

23 MARCH: 6.20am. Greater spotted woodpecker drumming. Chaffinches playing kiss-chase, all the day long in the trees.

Walking along the ride there is the tiniest mewing in a be-ivied sycamore. So I now know where the grey squirrels' new drey is. And they have kittens.

24 MARCH: Usually I only cry about dogs and horses. Even when I bade goodbye to my father, as the rollers took his coffin into the oven at Hereford Crematorium, I stood military straight in my covert coat, and merely inclined my head.

Just after 7pm my wife comes into the kitchen to find me sobbing my heart out. I cannot believe that tomorrow the boy who went to Hereford Library aged twelve and packed his briefcase with BB books will be the guest of honour at the BB Society AGM.

25 MARCH: The BB Society AGM at Sudborough, Northamptonshire. Drive past BB's Round House, his last residence, into the village.

In the village hall, meet Gordon Wright and Ray 'Badger' Walker – men who actually knew BB. I begin my talk. 'I've won prizes for my books, I've been on the *Sunday Times* top ten bestseller chart, but to talk to you today is the greatest honour of my professional life.'

And I mean it.

Some days later, I receive a mysterious flat package in the post; it's from Gordon Wright, the secretary of the BB Society. Inside the cardboard is the only print of BB's painting 'When Rufus Came to Stay', featuring a fox in a snowy wood in low winter light.

By the strangest coincidence the view through spruce in 'When Rufus Came to Stay' is exactly the same – the angle of the land, the point of penetration of the morning sun – as that in Cockshutt.

Rufus is the fox in *The Wild Lone*, perhaps the greatest of all BB's books. I gape at the painting for hours, trying to pinpoint what exactly in the scene is so mesmerizing. Is it the bluey exactness of the dawn snow? The ember-warmth of the pelt? Then I see it: the fox is on point, almost weightless, in just the manner of the alert Reynard.

26 MARCH: Sycamore leaves out, golden saxifrage growing in the ditch. New fern fronds push up through the earth; they are known as 'croziers', 'bishop's crooks', 'fiddle heads', although what the furled heads most closely resemble is a horse's head when rearing.

Blackthorn crystals crack open. Blackthorn is a flower of frost, a late taste of winter. There is no softness in it, unlike the bloom of hawthorn which is heavy with summer's promise.

*

A storm in the night; I stand amid the crashing trees, getting off on the fury. Nietzsche considered the urge to destroy a creative urge. The wind screams like a vixen on heat.

27 MARCH: On the leaves of the bramble bushes there are lightning zigzags; the trail left by the caterpillar of a bramble leaf miner moth (*Stigmella aurella*). The moth lays its egg inside the leaf, where the larva hatches and slowly eats its way through a tunnel just below the leaf's outer skin.

The larva overwinters in its 'mine', and emerges in the spring, silver, sheeny, and no bigger than a midge.

I lay my collection of mined leaves, with their individual white serpent trails, on an A4 piece of white paper; it should be an exhibition in a gallery.

In the afternoon: cut back the crack willow around the pool; the fluffy flowers fall on the water. Locally, the flowers are known as 'goslings'.

Water. It has no motion of its own; it is the mechanic betrayer of other forces. The breeze gently drifts the goslings to the far shore.

The jackdaws of a neighbouring wood fly up and mob a pair of red kites.

Last noise of the day: the clockwork whirr of a cock pheasant trumpeting, flapping its wings.

An unexpected item in the pigs' water trough. A toad.

I gaze hard into its primeval eyes. Toads are utterly inscrutable. There is no connection, as there is with a dog or a pig.

The first alder leaf appears. The Norse regarded March as the 'month of lengthening that wakes the alder'.

28 MARCH: Bramble buds; perpendicular to the stem. I pick them; they are good to eat as a wayfarer's snack. Useful too, I suggest, to the old Norse. March might have woken the alder, but it was also the time of enforced fasting, 'Lenct', as stores ran low. The Christian Church appropriated the word and the idea in 'Lent'.

The snakeskin buds of sycamore instantly catch the eye; when the buds unfurl, the leaves are languorous parasols, colonial umbrellas.

The sycamore is native to central and eastern Europe and western Asia. It was probably introduced into Britain by 1500, in the Tudor period, and was first recorded in the wild in 1632 in Kent. One telling proof of its interloper status is the dearth of native, local names. To my father, the sycamore was always the 'sick-of-more-work tree', because of its habit of dropping its large mucilaginous leaves on the rockery which then needed to be cleared way.

John Evelyn hated it equally:

> *The sycomor, or wild fig-tree, (falsly so called) is . . . one of the maples, and is much more in reputation for its shade than it deserves; for the honey-dew leaves, which fall early (like those of the ash) turn to mucilage and noxious insects, and putrifie with the first moisture of the season; so as they contaminate and mar our walks; and are therefore by my consent, to be banish'd from all curious gardens and avenues.*

However, sycamore makes good firewood (easy to saw, split with an axe, producing a hot flame), thus was it planted and coppiced in Cockshutt.

The johnny-come-lately tree has already achieved an ineradicable footnote in British history. Under a sycamore at Tolpuddle in Dorset, England, six agricultural labourers formed an early trades union in 1834. They were found to have breached the Incitement to Mutiny Act 1797 and were transported to Australia. The subsequent public outcry led to the release and return of the Tolpuddle Martyrs.

The Tolpuddle tree is cared for by the National Trust, who pollarded the tree in 2002 and again in 2014.

In Scotland, sycamores were a popular tree for

hangings, because their lower branches rarely broke under the strain.

At the entrance to the fox hole: brown chicken feathers from the poultry farm down the lane, which has wire walls six feet high. I have a sneaking regard for the sneak-thief that is *Vulpes vulpes*.

The chicken has been expertly, economically plucked.

Beyond the wood, in the fields, is the call-and-response of sheep, ewe to lamb and return. It is constant and rhythmic, in the way that waves gently lapping a harbour wall are constant and rhythmic.

29 MARCH: Evening, 8.15pm. Lichen light.

Robin sings with gusto, trying different refrains, experimenting. He is the philosophical songbird.

Hedgehogs now out of hibernation from their watertight nests of grass and moss. As I'm sitting in my chair one shuffles absentmindedly over my wellingtoned feet.

The wood is 'filling out'. There are no longer clear views through the trees. Gone is the sense of space, and light. The trees are crowding in.

31 MARCH: The floor of the wood around the stabiliz-ing tripods of the larch is potholed where the rabbits have scraped holes to get roots, these now white raw in the ground. Under the larch and the spruce I am sub-merged in silence as thick as water.

By 8.20pm Garway Hill is lost in darkness. I can smell the pheromones of spring.

The tranquillity of it all. The old Saxon line of settlement ran just west of me, beyond Garway was wild Wales. The difference still pertains.

The blackbird has finished her nest in the elm. The nest is a perfect bowl, of grass, straw and twigs, and plastered inside with mud. Years will pass, but the mud cup will last.

APRIL

Flowers

*The blackbird's gamble – Willow the pony – wood
sorrel – bluebells – a reading of bark – Edward Thomas
– what bird do you identify with? – the colour tide of the
woodland floor – wild garlic dolmades – fox cubs – the
leaf parade – transfiguration – alder gall mites*

1 APRIL: The buds, the leaves, the flowers, the blades
of grass; they increase so rapidly that I cannot number
them. Today the pool deafens with insect music, caused
by a million gauzy wings vibrating. The rising noise of
the insects signs spring as surely as the coming of the
cuckoo.

Two chiffchaffs in persistent competition. A cock
pheasant makes a lovely chicken *cluck-cluck* and a
noddy-headed walk.

2 APRIL: The blackbird in the elm lays her first egg;
as blue as the sky which looks down on it. The blackbird's nest is half open to the elements. Every species
takes a reproductive gamble; by being blue, and exposed
to light, the blackbird egg develops quickly . . . but is
within sight of predators.

Inside the shell, the speck which begins life.

3 APRIL: I go to the wood, briefly, but my mind is

elsewhere. Willow the Shetland pony has broken into the pig paddock and gorged himself on weaner pellets, causing grain overload. It sounds innocuous, but he has poisoned himself so severely he cannot move, and his eyes and mouth are blazing red. At 7.30pm Helen from the vet's arrives, looks, inserts a thermometer, listens to his heart. And pronounces: 'We should put him down.' She begins to shave a patch on his neck to administer the lethal injection, then suddenly says, 'Maybe we should try something else.' We, between us, half push and half lift Willow into a horse trailer on the yard, where Helen rigs up no fewer than two IV drips, then pours a gallon of bio-sponge mixture (to absorb poison) down Willow's neck. She sticks as many needles into him as a pin-cushion will hold. We agree that if his condition worsens overnight and he looks to be in pain, I will kill Willow with a shotgun blast to the back of the head. (There is a good reason farmers have shotguns: for mercy killings.) I sit beside Willow for most of the night, the Lincoln 12-bore on my lap.

4 APRIL: Helen visits 'the Little Man', as she calls Willow, in the morning, and at night. 'Well, he's no worse,' she says, cautiously optimistic.

5 APRIL: Willow's heart rate is nearly normal. He is

still in the danger zone, however, because of a possible secondary condition, laminitis, where sensitive tissues inside the hoof become inflamed. Again it seems relatively harmless, again it can be fatal. Helen makes little booties out of foam and gaffer tape for Willow's front hooves, to push the weight on to his back hooves, which are less likely to be affected.

Take a slasher to the nude, wiry bramble remnants left by the sheep, and still hung with tiny flags of their fleece. Build a bonfire in the wood, always a pagan delight, adding some spruce bits and pieces to the bramble tendrils; curls of smoke rise languidly through the oaks.

The first bluebells have chimed.

The tawnies have four chicks, and Old Brown is forced to fly by day to feed them.

7 APRIL: In the dingle there is an old toppled beech tree, which lies at full length, a shattered tube filled with the decay of its own substance, and tunnelled with beetles in shining armour. In the detritus the wood sorrel I planted two years ago is flowering wildly, so the prostrate beech has a second coming, as a gigantic ornamental planter.

*

Wood sorrel delights in such local names as cuckoo's meat, fox's meat, Easter-bells, cuckoo's bread and cheese, cuckoo's clover, Easter shamrock, sour dock, butter and eggs, bread and milk.

The plant was cultivated in earlier times; its lemony leaves are edible, and give a zing to fish sauces. George Orwell, in his autobiographical guise as George Bowling in *Coming Up for Air*, recalled that for an Edwardian boy sorrel was 'good with bread and butter' but 'sharp' on the tongue. Sorrel is from the Old French for sour.

In the southern parts of the country wood sorrel is out to greet the cuckoo and Easter, hence the multitude of country names citing the bird and the festival.

8 APRIL: When I was a boy we made expeditions in spring to Woolhope woods to bring back flowers. I remember my mother methodically searching a bluebell's petals with her fingernails. She had just been reading a book of Greek myths (Robert Graves' collection, I think) and was looking for an 'AIAI' mark on the flower. This she explained – she was a teacher – was the Ancient Greek for alas, said to be written by Apollo in memory of the death of the Spartan prince Hyakinthos. Since the bluebell was a member of the

hyacinth family, it had occurred to her that the bluebell might bear the stain.

The bluebell was unbranded, and it was only on consulting Grigson's *Flora* when we got home that we discovered why. The bluebell is *Hyacinthoides non-scripta*, the unusual specific part of the scientific name meaning 'unlettered', to distinguish it from the hyacinth proper. For me, it was already too late. An idea had lodged. The bluebell was the flower of Greek tragedy. It remains so, and breathing in the sweet, cool scent of the bluebells this evening I see the writhing prince in the mauve haze.

According to folklore, fairies were said to trap passers-by in bluebells. If you wore a wreath of bluebells you would be compelled to speak the truth.

Half the world's bluebells grow in Britain: they have played their part in our island story. When mashed, bluebells emit a sticky substance – the white gluey means by which, for centuries, pages in book spines and feathers on arrows were made to adhere.

Edge of the wood: there is a drifting blizzard of blossom from the wild cherry, or gean (hard G), four of them in Cockshutt. As the poet A. E. Housman noted, cherry blossom makes the tree appear to be 'hung with snow'.

Willow warblers, which arrived last week, are singing. The hawthorn, elder and sycamore are in leaf. A bumble bee lands on a goat willow catkin, soft bundle on soft bundle.

Sitting by the pool: overhead, the chatter of swallows passing north.

8.10pm. A walk around the wood with the .410; fourteen pigeons emergency evacuate from the larch.

Shooting wood pigeons requires observation of nature, quickness of wit and pure, distilled cunning. It's the way of the fox.

Pipistrelle bats, out of hibernation, hawk the pool. A casual sling-back moon.

A reading of bark: I've done this before, learned to identify trees by Braille of their bark. But Cockshutt has trees new to me, including the gean, with its pimply but precise hoops rising up the tree, to infinity. It is my abacus tree, the one I scratch when counting sheep, a hoop for a ewe. The spotty hoops are lenticels, raised pores that allow gas exchange between the atmosphere and the internal tissues.

9 APRIL: The anniversary of Edward Thomas' death at Arras, and I go into the wood to attend evensong. Me to the birds: 'Sing!'

*

Thomas was never decisive. Frost's globally renowned poem 'The Road Not Taken' is less an existential statement, more a joke about his friend Thomas's endless havering, in all things. While Thomas realized the necessity of fighting for England's fields, woods and streams he was nonetheless molested by irresolution. An owl, finally and irrevocably, convinced him to do his bit for King and Countryside:

> *All of the night was quite barred out except*
> *An owl's cry, a most melancholy cry*
>
> *Shaken out long and clear upon the hill,*
> *No merry note, nor cause of merriment,*
> *But one telling me plain what I escaped*
> *And others could not, that night, as in I went.*
>
> *And salted was my food, and my repose,*
> *Salted and sobered, too, by the bird's voice*
> *Speaking for all who lay under the stars,*
> *Soldiers and poor, unable to rejoice.*

This was no poet's posing. Thomas was so connected to nature that he considered his true countrymen to be the birds. Or the trees. When someone queried the meaning of his poem 'Aspens', about a stand of the trees beside a village crossroads, he replied, 'I am the aspen.'

*

Later: in the immense, pure quiet of the wood I hear the death of the vole, taken by Old Brown: a semiquaver of wail as it recognizes its anciently ordained fate, that the talons of archaeopteryx will some day puncture its back.

10 APRIL: The water of the pool locked into glassiness, utterly still except for the travelling silver wave of the moorhen's progress, made from evening light.

What bird do you identify with? It is a psychology test to outdo Rorschach. On this day I plump for the moorhen, the shy observer at the edge. You'll always find me in the kitchen at parties.

Why, why are people so enraptured with raptors?

By the pool a robin sings his heart out; further away blackbirds, and pigeons in pairs. In the dissolving distance a wood warbler.

11 APRIL: During the Second World War John Stewart Collis cleared a fourteen-acre ash wood in Dorset ('between Iwerne Minster and Tarrant Gunville'), neglected for eighteen years, where he could not 'advance a single yard unimpeded', not least for the 'throttling ropes of that hangman's noose called honeysuckle'.

He found that the woodland spectacle 'which fascinated me most, and encouraged me most, was decomposition'. The Renaissance had a motto *Et in Arcadia ego* – even in Arcadia, Death is present. Lord Tennyson saw it too:

> The woods decay and fall,
> The vapours weep their burthen to the ground.

One of the pigs has escaped into the wood, and snouted out a rotten elder log; the moss-coated bark is still firm, but the inside is soft like moist sponge; the further my fingers probe and expose the labyrinth, the softer the wood is. There is a fairy castle of miniature white fungi, a mushroom Neuschwanstein, and then a brown centipede, and its two mini-me babies; it is impossible to think of centipedes as babies. And two woodlice.

11 APRIL: The immutable, tidal waves of floor colour in an English wood: Green (dog's mercury)>Yellow (constellations of lesser celandines)>White (wood anemones)>Mauve (bluebells).

Sitting in my chair: the wind moves the goslings, the goat willow catkins, around the pool in a swirling Regency quadrille.

Odd how willows have become associated with sadness and mourning. In Shakespeare's *Hamlet*, Ophelia drowns near a willow tree. In biblical times willows were seen as trees of celebration.

The wood burns well too.

Up in the wild cherries there are posies and posies of blossom.

The hazel joins the leafy parade.

Against all odds, Willow the pony has fully recovered, and is officially signed off as a veterinary patient. This is not his first self-inflicted scrape with death. He is the pony with nine lives.

We love him.

13 APRIL: Wood pigeons mating; so preoccupied with their violent, neck-biting, dust-flying copulation in the larch that I walk to within three to four yards of them.

Since the trees are not totally dressed I can see the small birds working them; a big identifying advantage compared to when the trees are in full leaf, and birds are flitting shapes and soundbites only. The wood warblers have reached Cockshutt. Wood warblers are the last visitors from Africa, and the biggest of the 'leaf warblers' (the others being the willow warbler and the chiffchaff). Gilbert White knew the wood warbler's territorial song as a 'sibilant shivering noise in the tops of tall woods'.

The warblers are all carnivores, who come flying here for the spring hatch of insects. There are not many vegetarian birds in a wood.

And I wonder, as I do every year, how Darwin's cumbersome theory of evolution explains how the first warbler decided to fly five thousand miles to summer in Britain, then fly five thousand miles to Africa to summer there.

Four swallows arrive at Pool Farm in the valley, but move on. Migration of the birds is not a constant traffic but an eddying (as they stop for food, water) and a flow.

14 APRIL: First light. Dawn chorus in the wood begins with a pheasant, who stands in a lake of bluebells, the floral 'pride of the woods'.

Oak buds bursting, many of them extra plump because they contain both leaves and catkins. Ramsons (wild garlic) in the dingle stinking. The garlicky smell is potent enough to put hounds off the scent of a fox, should Reynard be cunning enough to run through a patch of them.

Locally common throughout Britain, ramsons like

shade, damp and neutral soil, hence its affinity for deciduous woodland and thickets. Superficially, the leaves of garlic are similar in appearance to the bluebell, but crush a leaf and if it smells of garlic then it is garlic.

The wood-cutting season might be done, but a wood can still provide. In the kitchen, the ramsons' leaves make a useful wrap, in the manner of vine leaves. Wild garlic is less fierce in flavour than the cultivated variety. The widespread dialect name ramsons is from the Old English *hramsa*. Ramsey in Essex and Ramsbottom in Lancashire are just two of the places which take their name from the locally prolific fruiting of the herb.

Wild garlic dolmades

80 wild garlic leaves	4 tbsp water
1 onion, minced	1 lemon
100g cooked risotto rice	1 tbsp tomato paste
2 tsp mint/water mint	1–2 cups vegetable stock
olive oil	

Blanch the wild garlic leaves by dipping into boiling water for 1 minute, strain and drain.

Sweat the onion in olive oil until translucent. Turn off the heat, but add the rice and all the remaining ingredients except for the vegetable stock, and mix well.

Put to one side.

Take 3 wild garlic leaves and lay side by side, so they are slightly overlapping. Put a pudding spoonful of mixture in the centre, and shape it into a tube running across the leaves. Now tightly roll the wild garlic leaves over the mixture – trying as you go to tuck the outer leaves over the mixture so that a small wrapped parcel is the result. When rolled, pin with a cocktail stick. If you find this method too fiddly, try laying the wild garlic leaves in a cross and wrapping this over a spoonful of mixture. The art of dolmades-stuffing is old and mysterious, and not learned in a minute. So you will need patience.

Continue making dolmades until all the mixture is used up. Using a large crock pot, or any pot with a lid, put the spare and discarded wild garlic leaves into the bottom as a lining. This is to stop the dolmades burning. Pack the dolmades in as tightly as possible, making double and even triple decks if necessary.

Spoon in three or so tablespoons of oil and the vegetable stock. Cook on a low heat in the oven for about 30 minutes, by which time the stock should have been sucked up.

The dolmades can be dished up with tzatziki as a starter, or placed on a bed of vegetables as a main course.

The above ingredients are all optional. Experiment at will. Halloumi cheese makes for good binding, while beef/lamb/pork mince is traditional.

16 APRIL: Snow falls through oaks, in end-of-the-world desolation. This is what the nuclear winter will look like.

Falling snow has an unexpected hiss.

17 APRIL: The clouds are great continents above my head, replicating the movement of the terrestrial continents over the Earth's crust in fast motion.

A wood pigeon perches on a twiglet of hawthorn. Perfect balance; but is it because the twig is stronger than I think, or the wood pigeon lighter?

On the pond a drake mallard takes off.

A nuthatch gives me a secret service *psst* as I pass.

Strings of greenish flowers dangle from sycamores. As an introduced plant, the sycamore has a relatively small associated insect fauna of about fifteen species. However, both male and female flowers produce abundant nectar, and the larvae of several species of moth use the leaves as a food source. The leaves additionally attract aphids, and then the hoverflies and ladybirds that feed on them.

18 APRIL: Oak leaves in limey translucence.

The geans, stolid straight poles, are flowerless until their tip exceeds adjacent trees of different breed.

The crab apple is the sweetest of all the trees, with its cups of pink-and-white upturned blossoms. The boughs drip, sculpturally, back down to earth. Today the two trees make bowers of pinky white.

Why are the tree flowers cream and white? So

pollinators under canopy gloom can sight them?

Under one of the firs the smashed shell of a pigeon egg looks like white cracked china.

The woodland air fades to TV-monotone grey. With all colour gone, there are only degrees of nuance, of insolidity. The wolf-light.

When I creep to twenty yards from the earth, I see that the four cubs are up above ground. Rolling. Tumbling. Chasing. They are lovely in the evening sun, their fur glowing with the spring light.

One night, I saw the vixen return with a rabbit; with true maternal consideration, she ripped the rabbit into four, giving each cub an equal share.

I try to edge closer still, but a traitorous tendril of bramble scrapes my trousers loudly enough to startle the cubs. They look directly at me; their eyes appear colourized with topaz. The cubs show no fear, but their mother is wise and calls them in from their play.

In April, the wood is the place to be.

19 APRIL: First day of low-level insect hum in the wood proper (as opposed to the pool); low level as in low down, and as in background.

The motion of the mallard on water: sliding magic.

23 APRIL: Bloody swarms of St Mark's fly in the wood, with their dangling legs, getting in the way. They are two days early, someone needs to tell them: St Mark's Day is 25 April.

24 APRIL: Evening going on nightfall, about 9pm. Sitting in my chair, tired from farmwork, bookwork, days which start at 5am. A hedgehog strolls past; I jump.

The buzzards are building a nest in an alder down in the dingle. One flies over with a stick in its mouth.

With a last gobful of mud from the pond margin, the house martin is blown back, rights herself, re-angles and flies on her predestined course: how does she know how to build? The mud house she was born in was made before her birth. Something in the DNA, or a secret coding in the universe the birds can tap into, and we cannot?

There is a point in descending darkness when trees are as substantial as vapour.

25 APRIL: Oak leaves out; beech leaves out. Beech, scientific name *Fagus sylvatica*, *Fagus* stemming from the Ancient Greek *phegos*, meaning 'edible', 'for eating', and this elegant tall tree that can reach over 120 feet also provides springtime lime-green leaves that may be popped into salads or soups. In the Second World War, Hitler's regime tried drying the leaves as a substitute for tobacco.

26 APRIL: Sunny. 10am. The floor of the wood becoming green, so Cockshutt is green below and above.

The ash remains stubbornly unleaved. 'Laggard to come, laggard to go' as the country saying has it.

Bird cherry by the pool in flower. The blossom has a distinctive smell; wet-dog lime.

A solitary peacock butterfly warms itself on the dried clay of the path. Sitting with its wings open, it reveals its four eye spots – suddenly displayed, they can frighten a bird away. Butterflies are easier to identify in the morning while they are warming up.

There is a nanosecond this day when a shaft of wisdom strikes: I am passing the beech, their sophisticated boles rising from brown ground, and I understand that they are a trope for another, alternative England. Different from the heartiness of oak.

27 APRIL: Snow, but light, nothing really. Spent two hours clearing sycamore saplings to halt the invasion of sycamore and extend the bottommost glade.

To kill a young animal – a puppy, kitten, piglet, foal – is a horror show; to take the life of an elderly beast is easier. It is the other way round with trees. I have no compunction about slashing the sycamore saplings to death, yet I – and likely you – would weep a little when felling a mature tree.

But with the rising sap and the nesting birds, this is the last wood work for a while.

In the afternoon: poke about in a dead tree. The holes and the cracks; I admire designs of bark, the twist in the bark of the sweet chestnut especially.

On the copper pond, the reeds are now two feet high; the more they grow, the smaller the pool becomes.

Along the ride the cock pheasant struts. The hen is on her nest in the bramble, which I came upon yesterday; she made a tiny movement of her head, which was enough to inform on her. She turned statue, and sat it out on her hot eggs.

28 APRIL: One of the sheep has fallen in the ditch below the pool; the ditch is about six feet deep. She has broken both of her front legs, so I fetch the shotgun.

Descending into the ditch is katabasis. The bottom of the ditch is an underworld of ferns, moss, slime, serpent ivy, the spine from some unrecognized animal. The coursing water booms off the walls.

Fresh blood has a surprising scarlet fluorescence, so vivid, so unlike the flat ochre of meat in the butcher's or on the supermarket shelf.

Perhaps one needs a climb down into the depths to appreciate springtime. Looking up, for the way out, a brimstone butterfly, just out of hibernation, jauntily flutters and glides across the sky ceiling. So wholly yellow is the male brimstone that he gave his name to all species of the insect: 'butterfly'.

29 APRIL: Transfiguration. Cockshutt is restless with spring, hormones, mating, nest-building. Growing. The spring life force is unstoppable – proved by the plants forcing through, pushing aside even the hard earth of the path.

The toing and froing of the birds: the woodpecker starts from a telegraph pole, squeaking like a rubber toy, and flies to the ash; the treecreeper is fast-mo back and forth across the end of the pool, etching an impression on the scene.

The air has a presence, a softness, a lightness that did not exist even a fortnight ago. Winter is leaving into memory.

At the top of the wood one of the cows is making a hell of a racket, trumpeting away. Then, from across the decades, the word bellocking plops into my head; it is the dialect term my grandfather would have used for the loud mooing of a cow.

This single word, a plastic credit card slid into a lock, opens a door. Dialect words from childhood rush to my tongue. The running of cows is *skelloping*; a badger is *Mr Teddy*, a sparrow *spadger*. I find I can compose half-sentences, clauses in dialect: I'm being *mithered* by a *jasper* (bothered by a wasp); it's *black over Charlie's mother's* (dark cloud harbinging a storm).

In a fit of nostalgia I make a list of old Hereford words appropriate to woodland, picked from Winifred Leeds' *Herefordshire Speech* and George Cornwall Lewis' *Old Herefordshire Words*:

BIRDS and ANIMALS

Billy-ploughboy – pied wagtail
Blue Isaac – dunnock
Bottle tit – long-tailed tit
Brock – badger
Bud-bird – bullfinch
Can-bottle/cannon-bottle – long-tailed tit (also for a skylark)
Chooky pig – woodlouse
Clover snapper – rabbit
Crane – heron
Devil's screecher – swift
Dishwasher – pied wagtail
Felt – fieldfare
Hickol – green woodpecker
Hoop – bullfinch

Kitty – whitethroat
Lady cow – ladybird
Maggoty pie – magpie
Mavis – song thrush
Mumruffin – long-tailed tit (also bottle tit)
Nettle creeper – lesser whitethroat
Pie-finch – chaffinch
Quist/queest – wood pigeon
Rail – moorhen
Richard – pheasant
Spadger – house sparrow
Stare – starling
Storm cock – mistle thrush (also thrustle)
Toby – fox
Writing lark – yellowhammer
Yaffil – green woodpecker

ARBOREAL

Apple-headed – of a tree where the branches are growing low
Arl, orl, orle – alder
Asp – aspen
Baby's bottle – wild arum
Bannut – walnut
Beechen – made of beech
Browse – hedge trimmings
Cag – stump of a branch
Chat – dead stick

149

Cobnut – hazelnut
Cockshutt – glade where woodcock were netted
Coppy – coppice
Daddock – dead wood
Devil's snuffball – puffball
Ellern, ellan – elderberry
Ellern blos – elderberry blossom
Ellum – elm
Elves' mittens – foxgloves
Fearn, vearn – bracken
Goslings – pussy willow
Mad Meg – bryony
Mauple – maple
Maybush – hawthorn
Oblionker tree – horse chestnut
Pank – knock down, especially apples
Pimrosen – primrose
Rundel – pollarded tree
Yimp – small twig

30 APRIL: Alder trees host the alder gall mite (*Eriophyes laevis*) which produces tiny, spherical blisters on the upper side of the leaves, green at first, but redder with age. Each blister has a narrow opening on the underside of the leaf through which the mite will leave when mature in the autumn.

The lower leaves of the paddock alder are poxy with

the galls, which is good news for the smaller insect-eating birds, who have picked so many galls the leaves now sieve light.

This is 'May Eve', when Shakespeare's *A Midsummer Night's Dream* (1594–5) takes place; the fairies' night.

MAY

Leaves

*Greenshade – tawny owl young – May Day festivities
– cuckoo – Nupend Wood – dawn chorus – mayflies
mating – a place to philosophize – the clustered bonnet
mushroom – lords and ladies – send in the aurochs – Oak
Apple Day – birdsong*

1 MAY: First proper day of greenshade. Buds and leaves bursting. The three tawny young are fluff balls in the ash, at first glance mistaken for rolled-up school socks. One falls to the crook of a lower branch; rescales the heights by climbing by the hook of his bill and the pleasingly homophonic crooks of his claws.

Mayflies on the pond drop and die like tragic ballerinas. For two years the mayfly lives a grub's life in the mud of the pond, then emerges for its brief hour of dancing glory, its hour on the stage.

Leaves have spiracles that allow air to get in; a leaf is made from sunlight and wind.

On my walk, I disturb the moorhen, which makes a low, leg-dangly flight over the water, scarring it.

A sudden yelp of greater spotted woodpecker. Some birds drop song; this woodpecker always sends its call along a lateral line, because it sings as it flies.

In the colour tide of the woodland floor: the bluebells are at their peak, their pomp, their majesty; the wood anemones' heads droop in acceptance. Only a single anemone is upright, a Canute flower.

Spruce are the enemy of the bluebell. The latter only grow under trees which let in light; even the larch and beech do this. I propose a 'euphoria' as the collective noun for bluebells. I float on mauve air going through them, and their delicate honey smell. Gerard Manley Hopkins looked into a bluebell and saw nothing less than 'the beauty of Our Lord'. Always the dab hand at neologism, Hopkins invented the word inscape for 'God's utterance of Himself outside Himself'.

A wood pigeon has fashioned a nest by the pig gate; such a jejune, sticks-every-which-way affair, my heart goes out to its maker, as a father's does to a child setting up home.

In the old time, people would go 'a-maying' today, and bring spring greenery into the home, a declaration that humans and nature should co-exist. John Stow, the Tudor antiquarian, recorded: 'In the moneth of May, namely on May day in the morning, every man (except impediment) would walke into the sweet meadowes and greene woods, there to rejoice their spirites with the beauty and savour of sweete flowers, and with the noyce of birdes, praising God in their kinde.'

May Day festivities often turned carnal. Sap rising, and all that. The anonymous pen behind 'A

Pleasant Countrey Maying Song', *c.* 1629, versed:

> *Thus the Robin and the Thrush,*
> *Musicke make in every bush.*
> *While they charme theor pretty notes*
> *Young men hurle up maidens cotes.*

2 MAY: There is a new sound in the wood, that of rain on full-fledged leaves; it is the shield-beating of Saxon warriors.

3 MAY: A new sight in the wood. Cuckoos are reclusive birds, but today at dawn-break one flies above my head, close enough for me to see the barred chest. It called (they sing on the wing) as it clipped through the beech grove. And I wonder: how much longer will the cuckoo be the signaller of the English spring? Since the early 1980s cuckoo numbers have dropped by 65 per cent.

Cuckoo. Cuckoo.

4 MAY: The cuckoo does not stay.

But, I thank God, the blackcap arrives, detected instantly by its signature *pebble-click* call, though I look and fail to see it.

Under the larch is a pheasant's egg, pear-shaped

and olive – but holed by the beak of a predator.

In the evening, eleven black swifts rush through the air: small, vociferous crossbows. The swifts are back from Africa, screaming their delight at being here. Welcome home.

The wood is filling with birdlife. Almost every hollow or rotten-limbed, ivy-covered tree has its nest of blue tits, woodpecker, chaffinch.

Collect tender oak leaves for oak-leaf wine. Leaves of the oak; quite wrought, ornate, for such a hard-case tree. Like the best of things, the oak has delicacy and strength.

I once dedicated a book to my son as my 'English oak' precisely because, as his eighty-five-year-old great-great-grandmother observed, holding his baby form, 'He's strong, but in a nice way.' She knew what she spoke of. The great-great-grandmother, Margaret, was a farmer's wife; more, she was herself the female tip of farm stock dating back directly, with no deviations, to the thirteenth century.

In the avian tower block there are five baby wrens.

6 MAY: I've a meeting in London, so start farmwork at 4.30am. Check the cattle and sheep. Give the lambs some 'creep' (wheat concentrate, largely to tame them). Load five sacks of sow rolls (again, wheat concentrate)

into the Land Rover, then drive across to Cockshutt, unload, fill the pig troughs. Unappreciative, the pigs snore on.

I look at my watch. I can risk ten minutes (anyway, I do not care: to be in the wood is my priority) and walk into Cockshutt.

The joy of dawn in a wood. Bushes are sculpted by imagination and the trickery of mist into creatures: a bear rearing, a tiger pouncing.

Sun rises to cancel the mist; in the forensic light the trees become reassuringly hard, solid. Friendly.

Evening, home again, and pass by Cockshutt: the trees revert to shadow, insubstantiality.

7 MAY: A busman's holiday, a visit to Nupend Wood, Fownhope, purchased for Herefordshire and Radnorshire Wildlife Trust in 1973 with money raised in memory of Dr A. W. Beech.

Nupend is the perfect wood; the template, the paradigm. I know it from childhood.

On this day of sun and rain, Penny and I take the Wye Valley Walk and swing into the top of the wood, where the conifers grow.

A red helium balloon, blown in from a party, is suspended from a branch and sucks the attention, in the exact same manner of the red buoy in J. M. W. Turner's 'Helvoetsluys'.

Nupend hangs either side of a limestone spine. Parts of the wood were once quarried, giving the affected ground a mesmerizing, convulsed aspect. The last time I saw earth like this was the war-pocked Hill 40, at Ypres. To nature, what difference the shovel or the artillery shell?

But this is mostly lovely olde woodland, principally of ash and oak, and then giant yews on the ridge, left stranded from pagan-time flood. The yew (*Taxus baccata*) is, with the Scots pine, our only native tree-conifer. Poor juniper is only a shrub.

I touch the yews' peculiar, time-smoothed trunks. On the fingertips they are sea-saturated driftwood.

Shall we be honest? Woods can be boring, with a claustrophobic, samey, wraparound view. The trees, the trees! Nupend raises the curtain on fresh scene-upon-scene, like in a play: towering oaks, meadow glades crammed with bluebells, firs used by thrushes as minarets, squinting slopes down to the River Wye. The earth sings notes of fox, fungus and fern.

Under spangles of faerie green-light there are wild strawberries. Pale clay paths radiate off, beckoning. Mud is embossed by deer hooves. Woodpeckers faintly tap the drum of the immense, aged silence. There are early purple orchids; the habitat supports stinking iris: we see its gladioli leaves by the side of the track. (Gladioli, aka *gladwyn*, Old English for sword.) The track back down to the car is pick-axed deep into the limestone, which forms escorts of crumbling slab walls. I push my fingers into the grey stone, and rip back time: two cockley fossils consecrate the place as truly ancient.

8 MAY: Dawn chorus in Cockshutt. One five-hundred-year-old folk verse proclaims, 'In summer when the shaws be sheen/And leaves be large and long/Full merry it is in fair forest/To hear the fowles song.'

But why do birds sing?

Orwell, in 1984, asked the same question, when Winston listens to a song thrush with Julia, somewhere in the Home Counties: 'For whom, for what, was that bird singing? No mate, no rival was watching it. What made it sit at the edge of that lonely wood and pour its music into nothingness? . . . But by degrees the flood of music drove all speculation out of his mind. It was as though it were a kind of liquid stuff that poured all over him and got mixed up with the sunlight that filtered through the leaves. He stopped thinking and merely felt.'

You see, Orwell was a naturalist. In 'Why I Write' he stated his list of loved things hadn't changed since childhood and these included 'the surface of the earth'.

Orwell was never really a socialist. He was a Tory anarchist, the label he himself gave to Swift. In 1984 the proof is there: Winston and Julia are free only in the deep English countryside, outside the reach of the Party and city socialism.

10 MAY: Raining: Spirograph circles on the pool. Fairy bubbles. Quite musical, with tings, liquid pings. The aniseed smell of ground elder.

The pool is a luxurious, oily-shimmery Rococo painting.

I find a broken mallard egg; the low music of breezed lapping water.

11 MAY: Dawn chorus II: starts with robin, then blackbird, song thrush, chiffchaff, willow warbler, wood warbler. They merge into a stream of song; I cannot distinguish their individual voices. They sing as one.

Emily Dickinson: 'I hope you love birds too. It is economical. It saves going to Heaven.'

Heat, which releases flies as if they have been thawed from ice. By 1pm very hot, bright, sunny. Butterfly time: peacock, cabbage white, meadow brown all on the wing. A green shield bug settles on a nettle, a blackfly settles on my piece of paper.

I gawp, actually gawp, at the fairyland beauty, the splashes of sun on the bluebells, all the tones of green.

A treecreeper, insect in beak, searches the sallow for another. Keeps flying over my head to its nest in the avian tower block. Where exactly? Takes me an hour to locate the spot. The old hobby of egg collecting at least required patience, skill, observation, time. The eggs, six, white with rust splatters.

Two red kites pass over; in the eclipse the wood coldens. Fantastically, when the kites have flown on, willow warblers sing in the willows. I throw a stick at a cat heading towards the moorhen nest – and raise a pheasant.

12 MAY: Beltane, in the Old Style calendar. The Reverend Francis Kilvert, of the parish of Clyro (which I can almost see from up a tree in Cockshutt), noted in his diary for 1870: 'This evening . . . I ought to have put some birch and wittan [mountain ash] over the door to keep out the "old witch". But I was too lazy to go out and get it. Let us hope the old witch will not come in during the night. The young witches are welcome.'

13 MAY: Dawn chorus III.
4.16am. Still dark, jackdaws over the hill *jack*.

4.19. After a stuttering *tic-tic* a robin, halfway up a hawthorn, performs full phrases.

4.22. Blackbird and wood pigeon greet the light.

4.30. Song thrush, willow warbler, blackcap.

5.00. Chaffinch, chiffchaff, great tit and wrens join in. Crescendo.

On average, grassland has seventy pairs of breeding birds per hundred acres, a wood four hundred. In a wood at 5am in May one can drown warmly in birdsong.

The song gradually drains away, leaving only a single stranded wood pigeon, and bluebells glistening in dawn's amniotic fluid.

In the evening: chiffchaff in alder, chaffinch in hazel. Both grate on my nerves with piercing two-note calls.

Bluebells: a dance troupe of ghosts in the evening light; wisps, distant when near. Some have seed heads forming. How quickly nature proceeds.

Cobbles of light and shade. The trees fatten every day; leaves take time to grow and swell, like the wings of a butterfly. The green of May trees sharpens the blue of the sky; after May, when the leaves are darker, the trees are complement rather than intensification.

Jew's ears shrivelled by the sun.

Gnats wurlitzer two feet above the pool, which is scummy.

An optimistic buzzard attacks four mallard walking through the wood. A gang of four magpies up to no good in the larch.

14 MAY: A drop in temperature, especially under the cover of the trees: always take a jumper to a wood till May is out. An orange tip butterfly colours the edge of the pool. Ditto the first red campion of the season. The rosebay willow herb beside the ride has reached a foot in height. Moorhens call from the security of the reeds, though the hen has taken up tenancy of the poolside briar patch for her nest. She shares the thorn fortress with rabbits.

The cold sun spotlights a dock leaf, to reveal it as skin, with arteries and veins.

A wind gets up, and the song thrush in the alder sings wild gobbled notes.

15 MAY: All the trees in full fig, except for the ash; the hawthorn 'heaped with may'. What a weight of leaves the trees support.

On the pool, mayflies mating on the wing.

The anemones gone, finished, as though they had never been.

In the cracks of an oak's skin, the caught red hairs of passing cattle. (Not the lustrous hairs of summer, but the rough fluff of the winter 'rug'.) A great tit descends for them.

17 MAY: In by the stile, past the so-elegant sweet chestnut, which is a piece of furniture as much as it is a tree. Then the imposingness of beech; a particular atmosphere, orientalist, minimalist. Beech has more presence than any other tree, even oak.

Golden tails hang from sycamores, sweetly fragrant, and tempting bees with their nectaries.

18 MAY: Sitting at the bottom of the Wishbone Oak, in the green shade; under a tree is always a place to philosophize. Ask the Buddha. Or John Stewart Collis: 'In the company of flowers we know happiness. In the company of trees we are able to think, they foster meditation. Trees are very intellectual. There is nowhere on earth we can think so well as in a thin wood resting against a tree.'

A thought: no one goes to a wood for conviviality. (No, not even lovers; they want privacy. In the lyric line of Keane, 'Somewhere only we know'.) A wood is a place for solitude, sanctuary. In a wood the only things you should meet are abundant nature and antique tranquillity, the present and the past tenses combined.

Up in the ash, a woodpecker taps away, as is his nature, and that of his forebears. In Cockshutt Wood. In Cockshutt Wood.

All nature has a feeling: woods, fields, brooks
Are life eternal: and in silence they
Speak happiness beyond the reach of books;
There's nothing mortal in them; their decay
Is the green life of change; to pass away
And come again in blooms revivified.
Its birth was heaven, eternal is its stay,
And with the sun and moon shall still abide
Beneath their day and night and heaven wide.

John Clare

19 MAY: Campions out; the gean still in white flower. From Orcop Hill, Cockshutt appears to contain chimneys billowing smoke. In a wood the gean grows in a lofty column, as opposed to the familiar bulb shape in the garden.

The oak puts up its high green roof. Primitive mythmakers conceived of trees as the link between earth and sky – and how right they were. Trees draw in water from the ground, and carbon dioxide and sunlight from the air.

I cut hawthorn, and hazel saplings as fodder for sheep. An ancient rite, hence the multiplicity of 'Spring Wood' around Britain.

Crab-apple trees: the two of them have bridesmaid's grace, with their long, flowery arms. The wild trees never fail to bloom and fruit.

Gloriously sunny, the low bright piercing sun through the canopy of dark trees. 'Shivelights' was Gerard Manley Hopkins' neologism for the sharp lances of sunshine through trees.

20 MAY: Kingcups; flowers gone, the seed heads formed, perfect medieval jesters' caps.

On a rotten oak half buried by time (equals: leaf litter) there is an eruption of *Mycena inclinata*, commonly known as the clustered bonnet or the oak-stump bonnet cap mushroom.

21 MAY: Rain: a squirrel heads down an ivied sycamore with a fledgling in its mouth.

The rain and the wind batter the ash leaves, so they shoal, pale belly up. The rain and the wind batter the pool, create shoaling pale waves.

23 MAY: A cuckoo calls in Hole Wood, once, once only. The resulting silence is as eloquent as any statistic. (In this entire spring in the Wormelow Hundred I hear two cuckoos; a century ago, my forebears would have heard a hundred.) Does anyone still write to *The Times* to claim the first cuckoo?

A glimpse of the fox cubs in the evening; their fur

changing from brown to the familiar red-orange, and their eye colour from blue to yellow. The vixen now takes them hunting by night. By October their adult life will have begun.

24 MAY: Lords and ladies coming into flower, a piece of woodland drama. A green sail emerges from the leaf, which unfolds to reveal an erect purple spike. As the season progresses this spike begins to stink like meat, and the tasty odour attracts tiny flies called owl midges. The midges fall through a ring of bristly hairs into the 'kettle' beneath the spike, and are trapped because of the hairs. In the kettle are the true male and female flowers. The midges bring pollen to the female flowers from other lords and ladies, and pick up pollen from the kettle's male flowers. Eventually the bristles wither and the midges escape to carry the pollen to the next lords and ladies. *Arum maculatum* was memorably described by Thomas Hardy in *Far from the Madding Crowd* as 'like an apoplectic saint in a niche of malachite'.

3pm. I scream down the M4 to Daunt's, Marylebone, to be interviewed by the novelist and nature writer Melissa Harrison, in front of a bookshop crowd. We get off to a good if funny start, since neither of us describes ourself as billed, as a 'nature writer', she because the term is so broad as to lack meaning, me because I consider myself a countryside writer. Me: 'My shtick is,

I think, obvious. I give the view of the countryside from someone who works there.'

25 MAY: The sun livens the insects; a greenfly lands on me. I go to snub it out, but what is its offence really? That it tickled? It is not reason enough any more.

There are now no views through the deciduous trees, due to the inflating of the greenery. Cockshutt is stacked with bird sound.

Of all the woods in all England, the migrant birds chose *this* wood, and they blessed it. And me.

Gobs of cuckoo spit, mostly on nettles but also on the brambles. Yellow flags in the reeds.

On the way out of the wood I perform the evening check of the pigs, still confined to the paddock.

I lean on the gate, real yokel style. Pigs have the big-bum, stiff-back-leg walk of hippos. Sly eyes, some of the boars. I'm with Walt Whitman:

> *I think I could turn and live with animals,*
> *they are so placid and self-contain'd,*
> *I stand and look at them long and long.*

26 MAY: Yesterday I ran six of our red poll through the top of the wood, around the oaks, partly to allow the cows some 'browse', partly so they would crap everywhere. How to increase invertebrate numbers in a wood? Send in the aurochs.

This evening a wood warbler flies off a perch in the oak to catch an insect rising from the cow dung. There are two pairs of wood warblers in Cockshutt, both in the oak grove.

The Victorian naturalist W. H. Hudson described their quivery song in the leaves as 'long and passionate . . . the woodland sound that is like no other'. They are a greener hue than the other leaf warblers, with a bright yellow beak.

28 MAY: 6pm. Sunny. At the sparrowhawk's perching post (a broken-off spruce, five feet tall), around the base a perfect circular necklace of blue tit feathers.

The pattern of leaves against the sky; the dottiness of silver birch; the clottiness of oak. To appreciate the pattern, one must lie on one's back under the tree and gaze skywards.

Sitting in my chair. A squirrel *chirr*s, tail upright, but flicks the tip cat-like in annoyance. I'm blocking the path, and it takes the long way round.

A wood mouse pitters about. BB decided mice look like gnomes, not without cause if you compare the feet of the two species.

Hawthorn blossom: facing west it is rusting, facing east it remains fresh as dew.

A woodpecker in alder, a goldcrest in conifers, the latter very loud. Lots of green cherries now, in wishbone pairs on the floor. Oh, and the tender air of twilight and lazy rooks.

29 MAY: Oak Apple Day; also Royal Oak Day, Shick-Shack Day – to mark the formal restoration of Charles II and also in memory of the oak tree at Boscobel in which he hid after the battle of Worcester in 1651 – hence the ubiquitous pub name, the Royal Oak. Oak Apple Day was popular here in the Royalist west (my mother's family apart, who were Cromwellian horse soldiers). Children wore oak leaves or oak apples in their lapels. Anyone not wearing one could be kicked, punched or even thrashed with nettles. Over time, Shick-Shack Day became a day of licentiousness, which was appropriately Royalist.

One of King Charles II's favourite snacks was blue violets, fried and eaten with sugar and lemons. He would be pleased, then, that they have flowered today. Country people called them 'blue mice', due to their shy peeking from under the spring greenery.

The male fern (its name, not its gender) is growing tall. It has abundant tall fronds, all shooting up in a green fountain.

Ferns are prehistoric jungle. Lie down in ferns and you can see dinosaurs.

Cockshutt also contains horsetail ferns, which women and men cleaned bowls with before Brillo pads and plastic brushes.

30 MAY: Already, I think, the summer diminution of birdsong. A willow warbler sits in an oak with the wreckage of ex-insects in its beak; it emits a soft *hoo-eet* call as it waits for me to pass, before it can fly down to its domed nest in the base of a bramble by the fence. Blackcaps, which nest higher in the brambles, make their *tac, tac* calls.

By the damp woodland path, the yellow pimpernel is out.

JUNE

Midsummer's Night

Elderflower, the summer flower of the wood – the pool, a
slow suicide – what is wood? – algae – the origin of trees
– 'tree hay' – A Tree Song – foxgloves

1 JUNE: At the dawn end of the pool the silver birches
are exactly reflected in the flat silver water; in the same
way that Jack playing cards have a mirror image. Where
did our moorhen chicks go? Lost them all to a neighbour's cat. Wish I'd been more of a sentinel, wish I'd
been more watchful.

Afternoon: warm, cloudy sun, and the muscatel perfume of the flaunt-it elderflowers. The English summer
begins with the elder's flowers, and ends with its berries.

Washing in dew gathered from elderflowers was
believed to preserve a woman's youthful beauty, and
concoctions of elderflower are used to this day in up-
market skin cleansers: Eau de Sureau. To this day
elderflower water, which is astringent, is listed in the
British Pharmacopoeia as a lotion for eye and skin
injuries.

The flowers also make that classic drink of the
summer, elderflower cordial. But I prefer to turn the
flowers into elderflower champagne, which precisely
captures in liquid form the oracular fizziness of the
flower heads.

Elderflower champagne

8 large elderflower heads	2 tbsp cider vinegar
4.5 litres cold water	750g white sugar
¼ cup wild rose petals (if possible)	champagne yeast
2 unwaxed lemons, sliced	a collection of clean plastic mineral-water bottles

Pick young florets, preferably in the morning when their banana aroma is at its strongest. Shake off the insects. The florets will not keep, so take them home as soon as you can, and in the kitchen 'fork off' (detach) any bits of stalk. This is bitter and will spoil your brew. If you can find them, wild rose petals add a subtle floral fragrance and pinkish hue.

Put the 4.5 litres of water into a large saucepan, together with the elderflower heads, wild rose petals, the sliced lemon and cider vinegar. Add the sugar and stir until dissolved. Sprinkle on the champagne yeast. Cover and leave to stand for 24 hours, remembering to stir twice with a wooden spoon.

Using a jug, bail the liquid through a sieve into the plastic drinks bottles. Put the tops on loosely and place the bottles on a plastic tray out of direct sunlight. Over the next fortnight, the champagne will ferment. When it has almost stopped, screw down the caps and store in a cool place. Allow a day or two for the fizz to build up, then the champagne will be ready for drinking – just refrigerate before pouring. The champagne will keep for months, but will gradually become drier and more alcoholic.

The joy of using plastic bottles is that you are more likely to avoid the calamity that sometimes strikes makers of elderflower champagne: glass bottles exploding from the pressure of the fermentation process. To check the pressure on plastic bottles, simply give them a quick squeeze. If the bottle is turgid, gently unscrew the cap until the gas hisses out, then retighten.

4 JUNE: End of day: drone of bees high in the trees; only the slightest flickering interruption in constancy, the aural equivalent of fluorescent light.

A mouse through the bramble; its *tic-tac* of feet amplified by the dry leaves.

Mosquitoes. And more mosquitoes.

Obviously, birds sing for territory, to prove their existence. I sing, therefore I am. That supreme caroller, the blackbird, is having an off-key moment, emitting Punch and Judy noises.

The weirdest event: a ring of barred feathers around the sparrowhawk's perch; something has killed the sparrowhawk, then plucked her. A fox?

A bitter happening on an evening when the surrounding hills are sweetly steeped in haze.

The reeds have their yellow flowers; the deadly nightshade is out.

5 JUNE: Cumulus cloud; the cock, hen pheasant and two chicks stroll around the paddock as if they own the place, freeloading on the pig food. Ripe, red cherries fall from the geans, so I scavenge them, the leftovers from the gorging squirrels and the birds. Even with a double extension ladder I cannot reach the highest, most Eve-tempting cherries.

A seductive red, the cherry, the scarlet of Hollywood lips.

6 JUNE: On a pool, panic has a pattern. The mallard drake takes off, into the hint of breeze, for lift. At the far end, the moorhen slinks leggily away through the reeds. She shows her white under-tail feathers as she goes.

Small wonder the waterfowl are alarmed. I have arrived at the pool driving a Kubota mini-digger, on caterpillar tracks, with the front loader clanking. I have brought the demonic din of the Industrial Revolution to their private Eden.

I turn the mini-digger's engine off. Rain finger-taps the roof, drips off, in through the open sides of the cab.

On the surface of the pool, the rain makes endless, repeating circles. Where gathered raindrops funnel off the trees, bubbles blow on the water, then pop.

From the security screen of green reeds, the moorhen *kerrupp*s at me. Her cry sounds as though her mouth is, rudely, full of water. Moorhens are misnamed. The bird has nothing to do with moors but everything to do with meres, meaning mires. Moorhens more than double the interest of a pond. A pond without moorhens is like a TV screen without a picture.

Heavier rain now, which makes strange *plurp*ing music on the water. The pond takes the colour of the June storm to become a black tar pit.

It is wet, but June muggy.

Although the year is slipping towards the midsummer silence of the birds, this noon the birds are singing. Willow warbler. Chiffchaff. Blackbird.

Another glimpse of the moorhen, scurrying along the grass on the eastern bank. With her red bill and yellow gangly legs, outsize shoes, she looks ridiculous, like a girl who has got at her mother's make-up box and wardrobe. The moorhen disappears into the base of the briar fortress where her nest is. The brambles are also tenanted by rabbits. Sometimes she hisses at them.

A cock chaffinch flies back and forth to his nest in the hazel, feeding gape-mouthed chicks with grubs. White grubs, green grubs.

Baubles of rain lie on the bent blades of the rushes beneath the digger's cab, and have silver, glam-rock sparkle. Shine on, you crazy diamonds.

Denys Watkins-Pitchford once wrote, 'A pool to me is as important as a house.' It's a view I share. When we have lived in houses without ponds, I have made them. A china Belfast sink in a cottage garden was a particular triumph.

Our current pool is industrial sized, about a third of an acre, and was double-purposed as fishpool and cattle watering hole. Mind you, with no fertilizers and agrochemicals to pollute them, almost all pools until the 1960s were fishponds, and teemed with eel, pike and trout. Some fat abboty carp linger in the pool, but for the last decades its prime purpose has been to provide drinking water for cattle.

The wood pool is a slow suicide; it fills with silt from ditches, with leaves from trees. To keep it alive requires vigilance.

Suddenly the rain stops. A pond changes by the hour, by the minute. Now it is placid glass, with only the very faintest scuff from the warm air. Insects emerge from their hiding. A blue damselfly, the smaller relative of the dragonfly, hovers above the reeds; an ancient insect atop a plant that thrived in dinosaur mud. Damselflies are among the oldest of all insects; fossilized remains found in coal strata reveal that enormous dragonflies, possessing wingspans in excess of twenty-seven inches, patrolled the swamps of the Carboniferous period.

The damselfly flickers away, a horizontal shaft of neon light.

The aniseed aroma of ground elder fills the air. Purple loosestrife bows in the heat. The peace is heavenly.

The pool returns to its innate monkish tranquillity.

The old cock chaffinch is still about his labour. He is the Protestant work ethic on wings. I should be about my job too. The reason I am here with the digger is that a pine tree has toppled into the pond, blocking the outlet. The tree needs to be lifted away. I switch the engine on.

I suppose I gazed at the pond for no more than twenty minutes. A kind of thirst was slaked.

8 JUNE: After two days of rain and wind, and broken branches, explosive bursts of warbler song, yet cacophonic, as if a hundred wet fingers were rubbed on glass.

I follow the air trail of a great tit to its nest, a hole in the high bough of one of the elm skeletons in the dingle.

Met a hedgehog on the ride: tipped my hat, said hello.

10 JUNE: The swifts swirl the scent of the elderflowers through the evening air. A shower of rain on warm earth; the woodland floor is braided with violets.

11 JUNE: Brambles in flower. We take dark shelter from the rain, me and the wood's animals both, in the grottoes of the lower branches.

13 JUNE: The hirondelles are stacked, airliner style, in the high blue: house martins, above them swallows, and topmost the swifts.

One of the joys of birds is that they cause us to look up, and think about matters spiritual.

14 JUNE: Windy, and the willows dance the dance of a million veils.

Taking half an hour off from shearing sheep, the hardest, most back-breaking work known to man.

Sitting in my chair: I am mesmerized by the lichen patches on a young(ish) ash, floating lumps in a 1970s lava lamp.

15 JUNE: Flying over the pond the red kite casts its reflection on the water. I have reflections of my own:

What is wood? On the inside of the tree is xylem, a mass of tubes that carry water with dissolved minerals up from the roots to the leaves. In broadleaves most of the xylem tubes are open all the way along but in conifers they are interrupted by perforated plates; the second group of conducting tissues form the phloem: strings of cells that carry the products of photosynthesis from the leaves downwards and outwards to the rest of the plant. The tissues of the phloem are on the outside.

Collectively the phloem forms a cylinder, enclosing the solid column of xylem within.

Algae on the surface of the pool, though any tree lover should hesitate before condemning the greenstuff.

Algae ventured on to land around 450 million years ago, and from these eukaryotic cells containing chloroplasts came, in the late Silurian period, the first 'vascular plants' – plants with plumbing systems. These early vascular plants invented lignin, which toughens cell walls. (Plants *sans* lignin are called 'herbs'.) Lignin is what turns wibbly cellulose into wood. A formula: wood = architecture + lignin. The first trees appeared in the Carboniferous period, about 360 million years ago.

The green algae on the pool, the overlooking trees, are the past and the present side by side.

The wood's leaves are turning a darker shade of green, so it is time to make 'tree hay'.

A discourse on the making of tree hay.

Collecting tree leaves for feeding livestock, usually from pollards, is a vanishingly small component of farming, though once it was widespread across Europe.

There is evidence that the practice pre-dates the making of hay from meadows, meaning it has been going on for three millennia.

Like meadow hay, leaf fodder or tree hay was stored for feeding to stock during the winter; food for the farm beasts was so lacking that, commonly, one in five animals did not survive winter, and survivors were often so weak they needed to be carried to pasture. Tree hay made a difference. Tree hay was also vital in drought; trees, with their deeper root systems and mycorrhizal fungal associations, can access moisture and nutrients and produce green leaves when the grasses have dried up. More, tree leaves are known to have medicinal benefits and stock will self-medicate where they have the opportunity.

Tree hay is produced by the cutting or breaking of limbs and twigs of deciduous trees and shrubs in full leaf. I do not have pollards, so prune lower limbs of hazel, beech, sallow, hawthorn, blackthorn, elm on trees and bushes daubed with a dot of white paint; only the trees and bushes I know categorically do not have birds nesting below twenty feet.

Tree hay is peasant work. The pruning, with a bill-hook or long-handled pruners, brings down the flies, and other flies, lured by the honey sweat of human, come in, bite, leaving on my neck a choker of shining pustules.

Entomologists use a 'beating tray' to collect and

study the invertebrate inhabitants of woodland. This consists of a light-coloured piece of strong fabric, such as calico, stretched across a frame.

I am wearing a white shirt, which serves just as well. My shirt crawls with caterpillars, aphids, flies, spider.

Half a day of cutting produces about a ton of tree hay. Another half a day to load it on the trailer of the Ferguson TEF tractor reversed up the ride. Taken to the barn at home, the branches are tied in tight armfuls with baler twine, and laid in a stack in a corner four feet deep.

Some of the tree hay from last year is still green, and good.

Thomas Tusser (1524–80), English farmer and author of the instructional poem 'Five Hundred Points of Good Husbandry', advised the lopping of 'all manner of trees', but ash, oak and elm (before its epochal demise) were the preferred species for tree hay. Regular re-cutting of elm can keep its bark thickness to a minimum and thus useless to the bark beetle, meaning tree hay aids the preservation of the elm, if in its diminutive form.

Wood fodder is hardly remembered. Indeed, I know stockmen who believe, with religious certitude, that the cattle and the sheep are best kept away from the hedge, the linear woodland, let alone a proper wood.

Of course, they have a point. There are animals on farms these days, mentioning no continental, commercial names such as Simmental or Texel, who would be woolly-headed stupid when confronted with tree browse, an essential in their ancestors' diet.

17 JUNE: Beating sun; one of those days when one can only work in a wood in gawping, disbelieving wonder at the scale of the tall trees, the oak and the beech, and our insignificance when standing next to them.

Another day of tree haying. Tough work now, the leather rigger gloves so moist with sweat they keep sliding off.

The insects: they jet-zoom, zoom zoom, kiddy-play jet planes or F1 cars about my head.

Endless flies: hoverflies, damselflies, mayflies, greenflies, blackflies.

Another half-ton done.

The highlight: I am so covered in tads, dribs, bits, frags of tree and bush that I pass for them. A gentle, mistaken blackcap perches on me.

And it was worth it, all of it, for that single moment.

I get lost in the job, as you do in physical labour, so the rest of the day is glimpsed specimens behind glass: a single puffball, eaten into a jack-o'-lantern by slugs; the movement of squirrels, causing an unseasonal crash of branches; sticky sycamore leaves; the new plantation of burdock by the stile, where none existed last year, the responsible seeds sure as hell planted by the pigs; four separate bluey clumps of rabbit fur; bluebells in their seed caps; my disappointment that we have no jay this year, a tick-box woodland stalwart (I love a jay's scream through winter wood); the cackling of magpies in the cherry tops; and under the spruce the lightest snuggled, doggy depression where the fox lies on the dry detritus there, red on red. I have seen him.

The speckled woods are about. They are unique among butterflies in that they can spend the winter as either a caterpillar or a chrysalis. The male is fiercely territorial and easily roused by any pretender that happens upon his turf: when that occurs, the rivals spin up through the woodland understorey in fluttery combat, until the interloper is driven off.

18 JUNE: A development: the moorhen female has moved to the island, using the bottom branch of an alder as a sideways slipway into the water.

The dappled shade of birch on the dapples of the fallow deer, five of them. I have never realized before that the whitey daubs on the deer's back are camouflage. Unfortunately, I instinctively slap a horsefly biting me, and the deer vanish into the thick, close air.

*

In the Middle Ages the birch became a lucky tree. Crucifixes carved from birch wood were hung around the necks of livestock to ward off satanic enchantment.

19 JUNE: The woodland edge; the ferns, the brambles, the grass, the flowers, the rambling rose, the honeysuckle. The demarcation between wood and edge is exact; one can, babyishly, step in, step out of the two worlds. There is no blending, only the absolutes of light and dark.

Up towards Garway Hill, a lordly peacock crows, sound of another age.

The honeysuckle bathes in the sun of the field, before winding clockwise around a hazel, deforming it. The purple loosestrife on the ride is out; in the ash a magpie hops up, peering and up to no good; he is the woodland's Child-Catcher.

One ash tree, just one, jiggles its leaves, making me wonder if it is ill, or perhaps ecstatic.

At 7.21pm: the arrow-thwack of a swift flying past my head.

21 JUNE: Birds heard at midnight at midsummer in the wood:

Old Brown, the tawny owl, intermittently.

A disturbed pheasant *cok-cok*s then dozes again.

A willow warbler.

Robin, who is the king of the night-singers, serenades me from the silver birch.

First light. Blackbird, wren, jackdaw.

All night shrews *pitter-patter*.

Moths clutter my face.

The awake, and the asleep.

Just before dawn the badger ambles to the pond, tortoise-extends his head and laps at the water. In motion badgers are shambolic rustics; in repose they are slinky expressionists.

From the three-foot reeds, the moorhen snipes at old Mr Teddy.

A Tree Song

Of all the trees that grow so fair,
    Old England to adorn,
Greater is none beneath the sun,
    Than Oak, and Ash, and Thorn.
Sing Oak, and Ash, and Thorn, good sirs,
    (All of a Midsummer morn!)
Surely we sing of no little thing,
    In Oak, and Ash, and Thorn!

191

Oak of the Clay lived many a day,
  Or ever Aeneas began.
Ash of the Loam was a Lady at home,
  When Brut was an outlaw man.
Thorn of the Down saw New Troy Town
  (From which was London born);
Witness hereby the ancientry
  Of Oak, and Ash, and Thorn!

Yew that is old in churchyard-mould,
  He breedeth a mighty bow.
Alder for shoes do wise men choose,
  And beech for cups also.
But when ye have killed, and your bowl is spilled,
  And your shoes are clean outworn,
Back ye must speed for all that ye need,
  To Oak, and Ash, and Thorn!

Ellum she hateth mankind, and waiteth
  Till every gust be laid,
To drop a limb on the head of him
  That any way trusts her shade.
But whether a lad be sober or sad,
  Or mellow with wine from the horn,
He will take no wrong when he lieth along
  'Neath Oak, and Ash, and Thorn!

Oh, do not tell the priest our plight,
    Or he would call it a sin;
But – we have been out in the woods all night,
    A-conjuring Summer in!
And we bring you good news by word of mouth –
    Good news for cattle and corn –
Now is the Sun come up from the south,
    With Oak, and Ash, and Thorn!

Sing Oak, and Ash, and Thorn, good sirs,
    (All of a Midsummer morn!)
England shall bide till Judgement Tide,
    By Oak, and Ash, and Thorn!

Rudyard Kipling

24 JUNE: Strange days indeed: a grass snake swims across the pool, in whiplash Ss. His emerald-green skin is foreign, fit for the tropics or a glass case, on the brown-water pool, but then he reaches the herbage of the reeds, when and where he fits perfectly.

We never become inured to the snake; he is always loathsome.

A golden drop of resin emits from a pine; next to the larch is a decent display of foxgloves, which the bees have claimed as their new best friends.

*

The foxglove is one of the most poisonous plants in flora, and yet its leaves contain a substance which is the source of one of the best-known and most widely used medicines in heart disease: digitalis. It was in the eighteenth century that William Withering, in Shropshire, investigated the folk-use of foxglove tea for curing dropsy (which is an accumulation of watery fluid in various body tissues and cavities). He guessed that it might have wider medical uses, and by 1799 – the year of his death – it was recognized as a valuable medicine in the treatment of heart complaints. There is a foxglove carved on Withering's tombstone in Edgbaston Old Church.

Strange days indeed. The nine o'clock shadows are long, but the fields glow from the inside, like the Habitat paper lanterns in my student digs in the 1980s.

Finally, the sun sinks below the rim of our world. Seeping through the wood, like friendly gas, is the fragrance of honeysuckle.

Honeysuckle gets its name from the sweet nectar in its corollas which makes them so good to suck. Pepys called it the 'trumpet flower', whose 'ivory bugles blow scent instead of sound'. In the Victorian language of flowers it stood for generous and devoted affection.

26 JUNE: Our wood glimpsed from up the lane: a line of green cumulus cloud, or a massif seen from afar. The oaks puff out, in green mushroom clouds; oaks demand light, and the sphere is the most efficient way to catch light.

28 JUNE: The summer wind, which rattles every one of the million million needles of spruce.

Hazelnuts are attired in their Regency bonnets. And then a curiosity: there are mullein caterpillars on the hazel, which is not listed as a larval food plant for the species. Feeling rather scientific, practically professorial, I take two caterpillars, and a handful of hazel leaves home in my hand, and put them in a jam jar. Attach a paper top, stab for air holes.

The mulleins, I can hereby certify for the record, do eat hazel leaves.

29 JUNE: Rain, plus rain. The clouds axe trees off at their necks.

The old log pile, which I made as a beetle-friendly habitat, has been ripped apart, the wood as rotten as tobacco. Whodunnit? A badger.

Already at midsummer the nuts are formed on the beeches; the cases are Desperate Dan stubbly; sycamore seeds hang in their 'horseshoes'; crab apples are berry-sized.

I prise open a bluebell seed head with my thumb-nail: each pouch has an average of fifty seeds; there are eight bells per stalk. So each plant has four hundred seeds to propagate itself. Multiply by the number of bluebells in Cockshutt. It is impossible not to feel nature's desire to maintain her bloodlines.

The rain brings up tiny mushrooms, grey pinheads, in the leaf litter beside the pool. Troops of them.

30 JUNE: Sprite light, and back in Cockshutt to one last load of tree hay.

The treecreeper has some leggy insect in its beak. Abandoning my billhook I peek in at her nest, in a socket where a limb has died off; the chicks are sightless, naked, hideous. All they want is the death of some mite so they themselves might live.

Such is nature.

JULY

In the Greenwood Tree

*Life in the top of an oak tree – tortrix caterpillars – oak
nut galls – Robin Hood in the greenwood – the 'insect
hour' – The wild perfume of honeysuckle – night in the
wood – moths – the woodland birds stop their singing –
what trees give the best shelter? – the redwood, the BFG
– the eye of a woodman – the killing of a squirrel*

3 JULY: It is said that more than four hundred types of
insects directly depend on the oak. I can believe it.

Up here in the Wishbone Oak, twenty-five feet
above the ground, I'm sitting legs astride a curving
bough. Holding on. A sailor up in the crow's nest of a
sailing ship in a galing sea would know the sensation.
You do not know a wood until you have lived in its
crown and looked down, as well as having stood on
its floor and wondered upwards, or hid behind a tree
and gazed upon deer . . .

The vertical world; the view down. Life at treetop
height; a wood is 3D – not a plane, like a field.

I climbed up the tree using my dextrous fingers,
which evolved precisely because our ancestors spent
80 million years up in the trees.

Something tar-like lands on my face, and in wiping
it away I brush a branch, and caterpillars rain into my
lap. Green oak tortrix caterpillars are 'loopers' – they
move by humping their back and bringing their tail
forward to head in a jerky progression. I swipe them off

my jeans and over into the abyss, but instead of falling they spin a silk lifeline, as fast as Spiderman does. For a moment twenty or more caterpillars are hanging by silk threads, then they start climbing back up. There is no getting rid of them.

More spots of black tar land on my face. I now realize what it is; it is 'frass', or caterpillar excrement. The leaves above me are heaving with caterpillars. What I thought was the hiss of breeze in high leaves is actually the sound of millions of caterpillars eating and defecating.

The average oak in Britain loses roughly half of its leaves each year to insects, with caterpillars sometimes taking almost all the first crop of spring leaves, whereupon the oak can respond with a second flush in May and June, known as 'Lammas growth'. (A misnomer, as Lammas, meaning 'loaf mass', is the Christian festival that falls on 1 August.)

The first half of the twentieth century saw the British countryside reach its maturity, based on the centuries of change needed for oaks to reach their veteran years. As the doyen of landscape historians, Oliver Rackham, pointed out, a single five-hundred-year oak is a whole ecosystem; ten thousand oaks aged two hundred years are not. We need trees to grow old, very old.

Since the Second World War perhaps a third of the ancient hardwood woodlands have been grubbed up, or built on for houses, roads, shops.

*

On the lower twigs of the Wishbone Oak, several oak nuts, galls caused by a tiny gall wasp, *Andricus kollari*. The wasp lays an egg in the leaf or twig, causing the oak to produce a chemical that forms a protective structure around the egg. The gall, russet-coloured, is nut-hard, and tough to crack. It takes a stone. Inside the gall there is a thick ring of dense honeycomb, and in a central hole, where a fruit stone should be, a larval, alien incubus.

4 JULY: We come over the Scottish border at dawn, like the raiders of old. Cumbria is a Tolkien shadow-land until Penrith, when the power of light separates clouds from hills, and hills resolve into steep, walled green fields.

My daughter, Freda, wakes up, programmed by an internal clock. The Lakes are one of her favourite landscapes. 'I suppose,' she says blearily, 'this is how most people see the countryside. Out of a car or train window.'

By Lancashire, she is asleep again. I jack up The Killers on the Saab's CD player. Hey, it's only middle-aged rock 'n' roll but it keeps me awake. Not till we are twenty miles from home, on the M50 beside the languid River Wye, does she next open her eyes. 'Of course,' she adds seamlessly to her comment from five hours previously, 'people don't see the work it takes to make the countryside look beautiful.'

Pulling up on the yard at home, she says, 'I like our road trips,' which is a divertingly lovely way of describing the eighteen-hour round trip I have undertaken for her in my guise as sole proprietor and driver of Dad's Taxi.

She goes into the house to pack for a holiday with friends in Cornwall. I drive to Cockshutt.

High summer in Herefordshire. Off in the oaks of Cockshutt, the pigeons call drowsily. On the tarmac a pile of horse poo steams with small silvery flies.

After checking the wheat I go into my familiar wood. The heat is stifling in the summer darkness. In a wood, this is the time of maximum light, maximum shade.

The wood and the world seem to hold their breath, and lines from Clare come to me: 'The breeze is stopt, the lazy bough/Hath not a leaf that dances now.'

The climax of the year. We have reached the top of the hill.

The writers of the old almanacs would have called this a dog day, after Sirius the Dog Star. Or maybe because of the heat which makes dogs, like the black Lab which accompanies me, pant.

I try to calculate how many hours in the last year alone I spent working in the wood. Two hundred? Then the year before that, and the year before that, and before that. Then all the hours other men have toiled here.

My mind could not take it in, all the hours of work down the ages to make the countryside look beautiful.

7 JULY: Under the beech, with its wide fans of branches, there is depth of shadow, where a man might hide. You think of Robin Hood in the greenwood, the summer wood; Robin Hood in the winter wood would have been a poor, persecuted creature. Robin Hood only exists in the summer greenwood of historical imagination, a place of hope, where the rich are robbed for the poor.

> *Who loves to lie with me,*
> *And turn his merry note*
> *Unto the sweet bird's throat,*
> *Come hither, come hither, come hither:*
> *Here shall he see*
> *No enemy*
> *But winter and rough weather.*
>
> *Who doth ambition shun,*
> *And loves to live i' the sun,*
> *Seeking the food he eats,*
> *And pleas'd with what he gets,*

> *Come hither, come hither, come hither:*
> *Here shall he see*
> *No enemy*
> *But winter and rough weather.*
>
> Shakespeare, *As You Like It*

Things fall down; the squirrel up in the wild cherry tree sends cherries and twigs crashing to the woodland floor. Cockshutt can accommodate a couple of squirrels. Six is too many. Something will have to be done.

10 JULY: Into the greenwood to check the glades, now wildly, brightly aflower with buttercups, clover, ox-eye daisies, the yellow and the red and the white, to determine when I should let in the cows for a graze. The answer: not yet. I need the ox-eye daisy (aka the 'sun daisy' in recognition of its joyous, bright yellow centre) to be going to seed, so the trample of cattle will plant more ox-eye daisies.

The morning is brightened by blood; the fox has left half a rabbit. Cherry stones on the woodland floor are bone balls, having gone through the digestive system of the birds.

Going into the wood in the evening, for five minutes 'me time' as a break from haymaking, but I am stupidly wearing shorts. It is 8pm, thus the insect hour, and my legs get, as we say in rural Herefordshire, 'bitten to buggery'.

11 JULY: How jaded are the leaves in the hedge. Already the cow parsley is tawny. The song thrush is on her second brood.

The wild perfume of honeysuckle, like other plants, is best in the evening, when one is relaxed and there is the touch of moisture to unlock smells.

Night in the wood: Old Brown calls; rabbits come to the entrance of their burrow, sniff, and eventually become bold.

How enjoyable the land is, when the sun has sunk below the rim of the known world, when other people have gone to bed, and there are stars over the dark, still oaks.

Moths come from every ferny corner, as white as fairies, emissaries from the court of the Midsummer Queen. The fern's lack of flower mystified country folk, who supposed the fern to have invisible seeds. If these could be harvested, it was believed the collector would be rendered invisible.

12 JULY: Entering the high summer wood is like entering a parish church; the same ancient silence, the same filtered, numinous light, the same smell of rotting wood.

The other way round, of course; entering a parish church is akin to entering a wood. The Gothic architects who made England's churches drew inspiration from the trees. Look at a deciduous tree bearing a heavy branch; under the branch, where it joins the trunk, you will see the tree has grown extra, supporting mass. A corbel. To stand under the high oaks is to stand in a nave, architecture imitating arboreal life.

Early Protestants, such as the Lollards, held services beneath the trees, and took trees as their pulpits.

13 JULY: This morning a swift was attacked mid-air by a sparrowhawk. It was damp, hence the swift was low. You can anthropomorphize swallows and house martins as cheery neighbours; swifts are remote, gym-muscular overlords.

In the wood at midday: swift-watching on a break from mowing hay. The swifts only hawk over the pool, and over the oaks, where the cattle were, thus confirming my hypothesis about the efficacy of cattle muck in increasing invertebrate life.

Pick up a feather – a buzzard's. The moorhen wanders along the far bank of the pool with her silly walk, as if jerked forward by her head, pushed by her tail.

The state of the trees: alder – the leaves green leather, incredibly strong, supple, and the cones formed; hazel – nuts formed but shell green and soft; hawthorn – draped with beaded necklaces of green haws; apple – in the space of a fortnight the fruit has become a red orb.

On the way out of the wood, the waist-high brambles are a mass of pink flower; the brambles smell grey
when crushed as I wade through to check the treecreeper's brood. Along the side of the ride, the rosebay
willow herb flames in the clearing (the first flowered on
22 June). During the last century the plant underwent a
genetic change which made it a more vigorous species.

About 9pm. And that particular summer stillness
when noises are single and precise; the cheep of nest-
lings; a train ten miles away, across golden fields; from
an irksome, gnatty thing, the whine of a dentist's drill.

9.45pm. Swifts still feeding.

10pm. Swifts stop feeding. There comes a time
when they can't see insects – they aim for their targets,
rather than trawl like whales for krill.

Swifts go to bed but their place in the sky is taken
by bats, which whip in, out and around the alder, and
circle the black glass pool, aerial sharks.

All the white moths come out.

A glimpse, no more than that, of badger at the
north end of the pool.

14 JULY: 3.30pm. The cries of four baby wrens from
behind the Tall Oak, in the Tarzan-vine honeysuckle.
Two nuthatches in the sycamores; the birds are always
somewhat bigger than one remembers. Brighter-hued
too – almost the kingfisher of the woods. I note that

the birds are as dextrous at climbing up/down, down/ up the sycamore bole.

The cep mushroom that erupted last week now dry and dying; old, varnished, cracked wood.

A cracked pigeon egg on the ground, from a birthing rather than a raiding by a predator.

At the sunny, warm edge of the wood, soldier beetles mating on cow parsley.

How dull life is inside a building; how clear and vibrant life is outside, in the world beyond the door.

15 JULY: The woodland birds have all but stopped their singing. Only the pigeons in the oaks call, and their drowsy *coo-cooing* serves only to make the afternoon more torpid. The birds are taking their summer recess, when they moult and become easy prey for the sparrowhawk. To sing would be suicidal self-advertising.

If trees have no brain, how do they remember? They *do* remember, they learn from experience. Two oak saplings I transplanted three years ago to the western edge of the wood are thicker by three millimetres than two I planted on the sheltered east of the ride. Wind-racked trees grow stouter than sheltered trees, light for light, water for water, like for like.

16 JULY: Early morning, and in the paddock below Cockshutt I catch the thieves of the pig feed in the act: the does are sweet in the sunlight, nibbling from the troughs; the buck stands under the sallow, every sinew ready to run.

About 10.15pm. Sitting in my chair: the perimeter of vision is fifteen yards; then the perimeter stalks me, closes in and swallows me.

Sometimes I pick up things as aide-memoires. Tonight a stickette to remind me that a bush has changed position, so it whipped me in the face.

The pale path. Who first made the path? Only in one place do I deviate from the path: the animals go under a low bough and I cannot limbo.

18 JULY: 27°C. Clouds are the gentle grey of a pigeon's breast, the prettiest creature the wood knows.

Sycamore leaves, sticky with honey dew. Sycamore seeds tight in their sacks, like a terrier's bollocks.

Sit in chair; a single wasp rasps a piece of fallen ash with its mandibles.

The pond ripples with the percussion of digger earthworks a mile away.

29 JULY: Sunlight in shards and splashes, then a cloudburst, allowing me to try a practical test to the question:

what tree gives the best shelter? I run around trying them all, and the answer . . . The beech is the umbrella tree.

30 JULY: Visit the Californian redwood, the BFG, have fun punching it, the outside bark being as soft as sponge. Higher up the tree, fragrant pine sap weeps where branches have been knocked off, wounds suppurating.

A slight breeze, but the wind-music of the woods is unaccompanied now by birdsong.

A dead rabbit is reanimated by the maggots inside it.

31 JULY: Peacock butterfly resting on a stone in the south glade.

Sitting in the crook of the Wishbone Oak, a station on my perambulation to check the readiness of the glades for grazing, I realize that I have turned over a new leaf in my life: I look at the oak – and the ash, the hazel, the elm, them all – with the eyes of the woodsman, the xylophile.

I translate trees into the objects they will make, the food they will provide, the shelter they will give.

Up in the distant canopy: a very slightly extra-bouncy branch. Sign of a squirrel.

Carrying a gun concentrates the naturalist's mind, sharpens the senses. I kill two fat adult squirrels with swinging left and right shots from the Lincoln 12-bore.

Aspens

All day and night, save winter, every weather,
Above the inn, the smithy, and the shop,
The aspens at the cross-roads talk together
Of rain, until their last leaves fall from the top.

Out of the blacksmith's cavern comes the ringing
Of hammer, shoe, and anvil; out of the inn
The clink, the hum, the roar, the random singing –
The sounds that for these fifty years have been.

The whisper of the aspens is not drowned,
And over lightless pane and footless road,
Empty as sky, with every other sound
Not ceasing, calls their ghosts from their abode,

A silent smithy, a silent inn, nor fails
In the bare moonlight or the thick-furred gloom,
In tempest or the night of nightingales,
To turn the cross-roads to a ghostly room.

And it would be the same were no house near.
Over all sorts of weather, men, and times,
Aspens must shake their leaves and men may hear
But need not listen, more than to my rhymes.

Whatever wind blows, while they and I have leaves
We cannot other than an aspen be
That ceaselessly, unreasonably grieves,
Or so men think who like a different tree.

Edward Thomas

This time last year: a copse beside the Argenton river in France, where every tree is an aspen. In breathless August, when all other trees are statues, the eighteen aspens quake. The scientific name of the aspen, *tremula*, is well conceived. They do, indeed, tremble, and this is due to botanical styling; the leaf stalks are flattened and bendy near the leaf blade, and the leaves (oil-cloth impermeable) are spaced apart, like signal pennants on a line. Individual leaves catch the wind, and flutter flexibly.

Trees are musical instruments. Each tree, like each human-fabricated musical instrument, is made different by design.

AUGUST

In the Green Shade

The month of lassitude – sycamore leaf rot – a stoat
'fascinates' – our English summer is shot – hunger
games – fallow deer at the pool – gathering blackberries –
woodlice – ash 'keys' – the airlessness of the high summer
wood – hauling tree trunks with a horse – the swift
departs – the fox cubs become independent

1 AUGUST: BB: 'August, of all the summer months, is perhaps the most dull and lifeless.'

Like BB, I dislike August, the month of lassitude.

All *seems* serenity: the heavy evening peace reaches out from the lawn to the rim of blue hills; we've got the hay safely gathered in, and the cattle went on the grass aftermath bang on the traditional date of 1 August, Lammas Day. The lambs are weaned, the ewes and rams sorted and checked, from teeth to trotters. I've a glass of Pimm's in my hand, a gorgeous valley in my eyes.

My neighbour's Land Rover thrums past on the lane, throwing up a pale dust cloud into the heat of the evening. On the lane-side hedge, the ladders of goosegrass are still in place, but grey with weariness and dirt. The cow parsley and the hogweed, dry and gone to seed, rattle in the humming air-wake of the Defender.

And you just know summer's lease is finished.

*

215

In the wood, the sycamore leaves are rotten with tar spot, caused by the fungus *Rhytisma acerinum*, and are falling in a private, precipitate autumn.

Summer woods: the darkness, the shadows, the unknown places, the mystery, the seclusion, the feel of being trapped. No clear run to freedom.

Summer woods shut out the sky; the comfort of light, by which we see, and steer.

Looking at my hayfield, with its succulent roll-in-me grass and bright flowers, next to Cockshutt: the wood is a brutal playground, with trees that scrape skin and sticks that trip feet.

2 AUGUST: I thought the 'fascinating' by a stoat, by which the little predator bewitches potential prey by dancing them a jig, was an old naturalist's tale but on this day I see something like it. (I should have been more generous-minded: both BB and W. H. Hudson write of stoats 'waltzing', and they are gospel.) A stoat is chasing its tail so fast it is a blurred red wheel. Two young rabbits outside their burrow below the holly are all eyes. The stoat stops and 'chatters', then spins again, but closer now to the rabbits.

When he breaks the dancing circle the next time he spots me, and bolts into the cowls of beech shadow.

Normally, when one enters upon a scene of animal behaviour one feels privileged. Yet the stoat's behaviour unsettles me, and it takes a moment or so to understand why. I feel I have fallen into a folk tale with anthropomorphic characters. The stoat's dance was wholly human in its reasoned guile, its intentioned mesmerizing of the rabbits.

6 AUGUST: A swooning August day, with a hot east wind from which there is no shelter. The wood is wearying with summer. The undergrowth is falling down, the leaves are old skin, and birdsong is feeble, except for robin's.

He is the uncontestable proof our English summer is shot. Now past his summer moult, the robin is already staking his winter territory.

I put the red poll cattle in Cockshutt yesterday, and they have the run of the place, except for half of the eastern glade, where I have put an electric fence to protect the harebells, blue, delicate and beautiful – and rucked by bees, fighting for their nectar.

Such is nature.

This evening the cows stand grouped around the Wishbone Oak, waving their tails against flies, to make the essential woodland portrait. They have torn into any ash leaves within browsing height; to borrow the phrase of John Evelyn, cattle 'are exceedingly licorish' after ash.

10 AUGUST: The hunger games. August, counterintuitively, is that awkward old month at the end of summer, when the majority of plants are past their profuse youthful best, the fruits of autumn are indigestibly immature, and the predator's easy-meat pickings – the very young, the very old, the babies, the slow – have already been snaffled, but many a predator still has clamouring mouths at home.

In summer, the mortality rate for the meat-eating birds and the beasts can be as great as winter. (I'm, improbably, an expert: I once spent a year living solely on the wild food that could be shot or foraged on our forty-acre Herefordshire hill farm.)

With the August dearth, I watch Old Brown's wife hopping around the wet glade pulling up earthworms, like an ungainly blackbird. The humble *Lumbricus terrestris* is currently the staple foodstuff for all sorts of animals. No fewer than seven buzzards hopped about on the aftermath of the hayfield yesterday, gulping the worms that had risen to the surface in the rain. All the buzzards' usual hawkish hauteur was dropped. They had the abandoned dignity of shoppers on Black Friday.

11 AUGUST: At the heat-drowned edge of the wood, the honeysuckle is in flower, but when I trace back the bine it goes ten yards into the interior, so that the roots are in the cool base of an ash. It's about 3°C cooler here than at the outer wall of trunks.

I'm standing, enjoying the welcome escape from the heat and the toil of fencing, when the fallow deer come tripping down through the green shade of the oaks.

As the five deer turn to stand on the bank of the pond their silk flanks catch the sun, so they shine with aureoles. In a line they stand drinking, almost inanimate, a plastic toy version of 'Fallow deer drinking at their watering hole'.

They must be thirsty to venture out of their fastness in the mid-afternoon.

Deer never relax. They are permanently aquiver, electric-wired.

I suppose some scent of me eventually carries on the warm air, because the young buck suddenly looks up. And bolts. The does are on his tail, no hesitation, no questioning. The five of them speed away, swerving through the obstacles of the trees, their bobtails flashing, waning white in the green gloom.

The deer breast the sea of briars, the emergency exit, out into the field. The days of sun have dried the wood's floor, and the percussion of the deers' hooves trembles the earth.

It is some sound, I tell you; it is the echo of the Norman *chasse* through the old wildwood.

When Vita Sackville-West wanted to invoke the unchanging pastoralism of England in *The Edwardians*, what natural items did she list? Well, 'the verdure of the trees, the hares and the deer'.

They made an odd group, though, the deer. Bucks usually spend the summer as solitary wanderers, or sometimes join together in small bachelor parties. And there were no fawns. I thought the deer were from Hole Wood, extending their range, but I am now thinking they are deer-farm escapees sticking together.

15 AUGUST: Up early, on the usual livestock round of pigs, cows, sheep. I walk in Cockshutt, intending to bid good morning to the moorhens, but get no further than the brush pile by the south glade, made when I cleared the trees there in our first year.

A badger has, overnight, tunnelled a drift mine into the pile. He can have only just departed, since there are woodlice spewing everywhere, attended by the piss-stink of their kind when they live in a large colony. (They excrete ammonia.) There are both common shiny woodlice and pill woodlice, as weird as each other. Woodlice are crustaceans adapted to terrestrial living; they need dampness because they still have gills for breathing. The gills are located on their legs. Eggs are ported around in a brood pouch which is kept moist, like a miniature aquarium. Pill woodlice can roll themselves into a tiny ball or 'pill'. In medieval England pill woodlice were swallowed live – as pills to cure digestive illnesses.

A woodlouse's closest relative is a crab. As I say, weird.

I half poke my nose into the badger's lice mine, but only half. Loitering are several *Dysderidae*, nocturnal venomous spiders with six eyes, who feed on wood-lice. Dysderids have fangs strong enough to penetrate human skin.

16 AUGUST: The ash trees are full of 'keys', their fruits, called keys because they hang in bunches reminis-cent of the lock picks of a medieval gaoler. Ash keys are widely and wildly rumoured to be pleasant eating when first ripe and peeled of their skin. I have never found them to be anything but screw-face astringent; raw, they taste of wormwood. And pickled they taste of wormwood. Those who love them say they taste like capers and are a piquant addition to cold cuts of meat and oily fish.

If you wish to pickle ash keys, pick fruits which are still green.

A handy tree, the ash, if one is confronted by an adder. According to folklore, adders so loathe the ash that they would rather flee through fire than through the leaves of *Fraxinus excelsior*.

17 AUGUST: A baby rabbit comes down to the pool and paddles in the shallow water before the smell of me drives it flying. I could see no reason for the playing in the water, except *joie de vivre*.

All this took a minute, if that. I was left listening to the music of the breeze in the reeds, and admiring the pretty patterns the breeze made on the water.

18 AUGUST: Walk around the wood at 5am, then get in the Saab with Freda to drive to the Somme, where we arrive at 5pm. I've a book about the First World War on the go, and need to check arcane aspects of the landscape. Farmers were told to diversify. I do books as my sideline.

The deciduous copses and woods of the Somme, in which British soldiers took shade from the Picardian sun, were simultaneously the earth and timber protectors of German soldiers.

Woods are nature's fortresses. The British 1916 Ordnance Survey maps suggest a count of forty-four woods and fifteen copses in the Somme battlefield, with Mametz Wood, extending to 186 acres of lime, oak, hornbeam, hazel and beech, the largest. High

Wood was known for the sweet chestnuts used to make pitchforks; after September 1916 it became renowned as a graveyard. It is estimated that ten thousand British and German soldiers still lie unrecovered within the bounds of High Wood. Some of the Somme's sylvan extents had open grassy rides – which gave excellent lines of fire to the grey-clad German defenders. And so the khaki casualties piled up at Mametz Wood and High Wood, names which to this day resound with the pity of it all. Those injured at Mametz Wood, amid the 'straggle tangled oak and flayed sheeny beech-bole, and fragile/Birch', included Private David Jones, author of the prose-poem *In Parenthesis*.

19 AUGUST: We leave Ibis hotel Albert (a hostelry so attuned to the Remembrance trade that the carpet has a poppy motif) at 8am and arrive home at 8pm. Then I go straight into Cockshutt Wood and am just in time to see a fox picking a blackberry in a late golden corner.

20 AUGUST: Along the woodland ride, and into the trees.

After the beating heat, the wood shade is pleasant ... The relief is short-lived. The trees' leaves are smothering cloth. Long gone is the springtime translucence, when every oak leaf was a pane of green glass.

The August woodland is dark. Low overhanging branches of sycamore form black grottoes.

Some extraordinary force has sucked all the oxygen out of Cockshutt Wood. The barrel chest of Willow, our Shetland pony, expands perceptibly. The air does not work; it provides no energy.

We gape, just as beached fish do.

I gee up Willow, with a shake of the long reins. We plod on, four-four beat, me walking behind him, a lonely wight in an English wood, going deeper into the murk. Willow, a palomino, luminesces; he is literally a guiding light.

August is summer's breathless zenith, when the vegetation has reached the limit of growth and the tree canopy is densest. In the wood, it is night in mid-morning.

The woodland silence is almost total. Blackcap, willow warbler, wood warbler all stopped their singing in mid-July. Even the chiffchaff has ceased his piercing two-tone chant.

But there is a rhythmic plague of flying insects. Lucky Willow is doused in lemon-astringent Naff Off fly repellent. I am not. Midges, mosquitoes grope at my face; I continuously bat the slate-grey horseflies away.

Up in the honeysuckle, which is still in flower, some bees drone on; a drone which dumbs the brain, and drugs the wood into stupefaction.

The path swings beside a glade, and there is a sudden joyful stab of sunlight. A peacock butterfly rests on a stump, drying its wings, after emerging from its chrysalid case in the nettle bed.

August is a quiet month in a wood, if a busy one. Large white butterflies ('cabbage whites' to all gardeners) are on their second brood; a new gatekeeper butterfly threads its course through the trees like a blown leaf.

Hedgehogs also produce second litters in August, while the first trot under cars on the lane, as they did last night, smearing it in scarlet and black. At least Cockshutt is at the edge of the local badger's range, so the wood's hedgehogs are not scooped apart by his mechanical grabbers.

Two long-tailed tits, with a late brood to feed, work over a silver birch branch, then start on the branch again. Eat. Repeat. A long-tailed tit takes an insect every 2.5 seconds. Or thereabouts.

August is a quiet month on a livestock farm too. We 'hayed' in July, and the maize is not cut until October. The lambs are weaned, the cows not due to calve until September.

So Willow and I are on DIY duties, off to haul two Norwegian spruce trunks out of the wood. The spruce is for poles to make a field shelter. In the Year of the Lord 2017 there are still times when horsepower is all; only an equine can fit along the woodland path.

On an alder a fallow deer stag has left shreds of

velvet. In August, when the new antlers are fully grown, stags are eager to rid themselves of the madly itching fabric. On hot days like these, when the urge to itch becomes unbearable, they rub their antlers against rough-barked trees until hard bone shows clean beneath.

Under a hazel, there is the shell detritus of an immature squirrel who has rushed through the hazels trying the nuts and finding them all still sour and whitish-green, immature themselves.

The crab apples are swollen with a sort of promise. They make the jelly for autumn game and Sunday morning toast. One of the beech trees is starting to show golden. Every year, it is the same: this one tree is the first to sign the dying of summer.

Somewhere at the north end of the wood, where it points to Hay Bluff, a wood pigeon clappers, finds a hole in the larch and escapes the oppressive heat. From the same place another pigeon calls: 'Coo-coo-coo/coo-coo-coo-coo/coo-coo-coo-coo/coo—' The ending is abrupt, as though the bird decided its soft summer melody hopeless, and gave up.

I tie the two trunks to Willow's harness; the blue nylon rope around the dead trees is electric vivid in the pall of the Norwegian spruce.

We start back. The trees close in. Birch. Alder. Sycamore. Hazel. Sallow. Beech.

We take the path alongside the pool, to try and

find some air and light. The tarry water is unmoving. In the reeds the moorhen calls her alarm. (After all this time, my beloved bird?) Her protest lessens but is never wholly abandoned, and she continues to bark intermittently at me from behind her fortress of green swords.

Suddenly, from out of the sun a swift dips down, takes a sip of water, and screams a last farewell. The 'devil's screecher' is migrating south. The stay-at-home moorhen is unimpressed, and continues her scolding.

I can tell you why the shade of the August wood dismays. It is a foretaste of winter's shadow.

23 AUGUST: 25°C. The moving leaves betray the invisible wind.

A five-minute trot through Cockshutt, trying to find a missing sheep, a species ordained a flock animal, and yet so bloody-mindedly prey to individualism at all the wrong times.

Scan the scene: ragwort; nightshade flowers; white marble clouds; a green dragonfly; honeysuckle flowers giving way to bright red berries.

But no stray sheep.

29 AUGUST: Hot, again. Wood gloom, as though the light is turned off. Lots of large white butterflies; their flight half determined, half dilettante.

Mosquitoes whine (a pleonasm, I grant you: their name comes from the Greek *muia*, an attempted rendering in a word of that irritating noise they make when flying). Apples silently fill and grow.

I see, for the first time, the autumnal, white-mist webs of spiders drenching the bramble.

30 AUGUST: Squirrel = tree rat, though you have to smile at the high-wire antics.

Swallow chatter; buzzard *pee-ow, pee-ow.*

Three hundred or more house martins mass on the barns. The flight of the birds . . . The departure south by the swallow tribe is the unmistakable, unbridgeable divide between summer and autumn.

But still you cling to your dreams of an Indian summer.

From down in the valley, across the harvested fields with their ziggurats of waiting straw bales, comes the chatter and clink of a drinks party. Distant voices, overexcited ambience.

Tonight, one of those farming happenstances that make you do a fist pump, raise a salute to the providential god. The lost sheep is waiting by the paddock gate on the lane.

Open the gate, and in the sheep goes, *baaa*ing,

happy as Larry the Lamb. If I was on Twitter I'd hashtag the day #Result

31 AUGUST: About 9pm, a young fox out by him/herself. The fox cubs have been sent out into the world to fend for themselves. The wild lone.

SEPTEMBER

The Birds Have Flown

Autumn smells as thick as curtain – glistening moss –
being in a leafy wood: experience in pointillism – put the
pigs on the beech mast – roasted beech nuts – chiffchaff
renews its song – holly blue butterflies – a weasel –
nutting – the yew in the churchyard of St Bartholomew's
– acorns falling – a hallucination of blewit mushrooms –
the health-giving properties of elderberries – greater
spotted woodpeckers chacking for territory – hedgehogs
building up fat reserves for winter

4 SEPTEMBER: A wet wood, hanging with autumn-rot smells as thick as curtain, but Cockshutt is still green in parts, especially the glistening moss on the stones and the boughs beside the trickle-ditch.

The ragwort is a shocking, inconceivable yellow under the spruce.

A fleeting glimpse of a rabbit – a flash – as it warps to shelter speed.

This is how it is in a wood with canopy: sights and sounds exist in bits. Experience in pointillism.

5 SEPTEMBER: In over the stile, and for the first time the beech mast crackles satisfyingly underfoot, the shells open in a three-way split, the kernels mahogany and shiny.

A gang of six chaffinches is sufficiently absorbed

in the epicurean delight of eating beech kernels that I almost tread on them before they fly off. Chaffinches, when working ground, seem to nod their heads; rather they keep their head still, the eyes on the prize, and jerk their body along to keep up with it.

With this much mast already fallen, and the finch time beginning, now is the moment to let the pigs gorge, and expel the cows (beech and acorn are poison to them), among them the girl calf born in spring and now already half the size of her ma. I spend the early afternoon putting up an electric fence, three strands of horizontal polywire on plastic poles around the dingle and the bottom of the wood, but protecting the glade. Snouting pigs are indiscriminate diggers.

The electric fence is attached to a tractor battery. I let the pigs into their new run. The scream of the weaner who puts her wet nose on the fence is not contained by the trees.

Beech trees do not crop every year, but every three or four years, when the harvest is likely to be heavy.

Some 20 per cent of the beech nut is comprised of a thick sweetish yellow oil excellent for frying.

To extract the oil, pulverize beech nuts in an electric grinder, or by the keep-fit equipment of a pestle and mortar, and squeeze the pulp through a sieve, muslin bag, or failing these a pair of washed tights. A kilo and a half of mast should give you about 250g of oil. The oil is best if you can be bothered to de-shell each nut, and even better if you scrape off the slightly astringent skin of the kernel. But this is a fiddly operation.

Beech oil was commercially produced in Britain until the middle of the Victorian era, and is still preferred in some French cooking circles over the ubiquitous olive oil.

Beech nuts are mildly toxic raw, so always cook them before eating. That said, they can be used neat to flavour gin.

Roasted beech nuts

Remove the kernels, place on a baking sheet and sprinkle with extra-virgin olive oil. Bake at 180°C/Gas Mark 4 until golden. Drain on kitchen towel, toss in ground sea salt, and eat as a party nibble.

8 SEPTEMBER: This, I suspect, is our last good day. The house martins are still here, and autumn has failed to sink its barbs into absolutely everything. Sunlight through hawthorn is surprisingly verdant, and has all the pretence of youth.

A dead wood pigeon on the path through the woods, with no marks of disease or harm from a predator. So,

I suppose some wood pigeon die a natural death, live their allotted lifespan. And I still maintain that the wood pigeon is our most underrated bird: it is good in the air, sparrowhawk fast; its sweet lowing makes a summer's day; it is good on the plate; and the rose-china blush of its breast is art.

9 SEPTEMBER: The day warms from a bad start, so by afternoon there are white butterflies and crane flies on the ride.

The chiffchaff renews its song, but it is less spirited than the full-throated performance of March. As Sir Edward, Viscount Grey, put it, the chiffchaff's autumn song is 'a subdued repetition . . . a sort of quiet farewell before the chiffchaff leaves us on its long journey southwards'. The chiffchaff is the last warbler singing; all the rest have departed into memory.

A brood of holly blue butterflies, so late I think them out of time, is on the wing, flying around the ivy, where they sup nectar from the green flower.

11 SEPTEMBER: Sitting in my chair, I am suddenly aware of being watched. A weasel is sitting up beneath the Tall Oak, so close I can see every whisker. The black-button eyes are utterly without gentlenesse.

Weasel. Say the word. There is no fluffy, Bill Oddie

way to pronounce 'weasel'. The animal is synonym for cunning, of deadliest sort.

The weasel is from the old time, yet it is also the perfect modern predator; a miniaturized, malevolent twelve inches of muscle with needle teeth. She is plainly unable to categorize the shape in the chair, with its stink of sheep (I have my farming coat on). Eventually she determines on discretion, and slides away.

12 SEPTEMBER: Hazelnuts sufficiently ripe to fall when the tree is shaken.

14 SEPTEMBER: Holy Cross Day (aka Holy Rood Day), a subject about which John Clare wrote to William Hone:

> On Holy rood day it is faithfully & confidently believed both by old & young people that the Devil goes a nutting on that day & I have heard many people affirm that they once thought it a tale till they ventured to the woods on that day when they smelt such a strong smell of brimstone as nearly stifled them before they could escape out again – & the cow boy to his great disappointment found that the Devil will not even let his black berrys alone & he believes them after that day to be poisoned by his touch.

William Cobbett, a decidedly earthier character than Clare, observed that 'a great nut year is a great bastard year' due to the ungodly fun folk had when out nutting.

15 SEPTEMBER: In the Middle Ages, the hazel was allocated to St Philibert (Filbert) because his commemorative day was 22 August, when the nuts were considered ripe. Neither I nor the squirrels nor the wood mice have ever found them properly ripe until mid-September. Since the devil, who is best avoided, went nutting yesterday, I go gathering hazelnuts today.

The modern English name for hazel derives from the Anglo-Saxon *haesel*, meaning hat, in reference to the frilly cap in which the nut sits.

The nomenclature might be medieval but the culinary use of hazelnuts is prehistoric, and they formed an important item in the diet of Mesolithic hunter-gatherers. Pick a hazelnut to eat and you are the first man, the first woman, the English Adam and Eve.

Hazelnuts have, proportionately, more protein than a hen's egg, as well as a significant amount of oil.

In the kitchen: put shelled hazelnuts on a baking tray (150–200°C for about ten minutes) and watch like a hawk because they will burn. When cool, I sprinkle some salt over them, and snack. Part of the consuming pleasure, I confess, is having beaten the squirrel to Cockshutt's bounty.

Hazelnut and mushroom pâté

Serves 4

1 small red onion, chopped
2 tbsp extra-virgin olive oil
1 clove garlic, crushed
150g porcini (cep) or chestnut
    mushrooms
1 tsp cognac/brandy
100g roasted hazelnuts

250g smoked tofu
1 tsp fresh rosemary, chopped
1 tsp fresh thyme, chopped
1 tsp shoyu
1 tbsp water
salt and freshly ground black
    pepper

Fry the onion in the olive oil until caramelized. Add the garlic and mushrooms and fry over medium heat until the mushrooms soften. Remove well away from heat and add the cognac or brandy.

Put the hazelnuts in a blender and blitz. Then add the tofu, herbs, shoyu and onion-mushroom mix from the pan and process until a firm paste emerges. You may need to add water. Season. Serve on toast.

16 September: Visit to St Bartholomew's, Much Marcle. In the graveyard there is a yew, 1,500 years old, its girth in 2006 measured at thirty feet eleven inches. One can sit inside the yew, on a three-sided bench; it is a rustic room.

Heartwood of yew has high crushing strength; yew sapwood has high tensile strength. The archers of medieval England made their killing longbows from the yew, from the part of the trunk where heartwood meets sapwood. With dark heartwood on the inside and the light sapwood on the outside, the yew bow gave tremendous spring, and made short work of French knights at Crécy and Agincourt, despite their expensive armour.

Robert Graves claims in *The White Goddess* that the yew was the symbolic 'death tree in all European countries'. As with St Bartholomew's, many an English churchyard has a yew tree that pre-dates it, suggesting the Christians engaged in a little light religious appropriation, building their churches on druid sites.

The yew contains highly toxic alkaloids in nearly all its parts, as befits the graveyard tree. (The one edible part is the berry, providing the seeds are not consumed.) Thomas Gray trades on the dark association in 'Elegy Written in a Country Churchyard':

> *Beneath those rugged elms, that yew-tree's shade,*
> *Where heaves the turf in many a mould'ring heap,*
> *Each in his narrow cell for ever laid,*
> *The rude forefathers of the hamlet sleep.*

And today, having sat in a yew tree, I see that Wordsworth's 'Lines Left upon a Seat in a Yew-Tree' are scripted in clear, distilled truth:

> *Nay, Traveller! rest. This lonely Yew-tree stands*
> *Far from all human dwelling: what if here*
> *No sparkling rivulet spread the verdant herb?*

What if the bee love not these barren boughs?
Yet, if the wind breathe soft, the curling waves,
That break against the shore, shall lull thy mind
By one soft impulse saved from vacancy.
        Who he was
That piled these stones and with the mossy sod
First covered, and here taught this aged Tree
With its dark arms to form a circling bower,
I well remember. – He was one who owned
No common soul. In youth by science nursed,
And led by nature into a wild scene
Of lofty hopes, he to the world went forth
A favoured Being, knowing no desire
Which genius did not hallow; 'gainst the taint
Of dissolute tongues, and jealousy, and hate,
And scorn, – against all enemies prepared,
All but neglect. The world, for so it thought,
Owed him no service; wherefore he at once
With indignation turned himself away,
And with the food of pride sustained his soul
In solitude. – Stranger! these gloomy boughs
Had charms for him; and here he loved to sit,
His only visitants a straggling sheep,
The stone-chat, or the glancing sand-piper:
And on these barren rocks, with fern and heath,
And juniper and thistle, sprinkled o'er,
Fixing his downcast eye, he many an hour
A morbid pleasure nourished, tracing here

An emblem of his own unfruitful life:
And, lifting up his head, he then would gaze
On the more distant scene, – how lovely 'tis
Thou seest, – and he would gaze till it became
Far lovelier, and his heart could not sustain
The beauty, still more beauteous! Nor, that time,
When nature had subdued him to herself,
Would he forget those Beings to whose minds,
Warm from the labours of benevolence,
The world, and human life, appeared a scene
Of kindred loveliness: then he would sigh,
Inly disturbed, to think that others felt
What he must never feel: and so, lost Man!
On visionary views would fancy feed,
Till his eye streamed with tears. In this deep vale
He died, – this seat his only monument . . .

William Wordsworth

Wordsworth might have added that the female yew, feathered and black as crow, with a red berry, is exotic in this English sky.

19 SEPTEMBER: Sweet chestnut leaves hang, smoked-kipper brown and shaped; cep mushrooms rise from the earth under pines, plus their bulbous 'babies'; squirrel chitter and chatter in the oak.

All trees on the turn; the earth smells of musk.

Fallow deer barking over in Hole Wood, and so the rutting season begins.

20 SEPTEMBER: Hot. Tired acorns plop to the ground. Blackberries now the general foodstuff of many a bird and animal. A shrew pauses on its hind legs to sniff the summer air with its twitchy snout. Its incessant movements in search of food earned the creature its name, derived from the Old English *screawa*, to swirl. Shrews are annuals, they last no more than a year.

This shrew, with his jaded fur, is entering the winter of his life. His children will continue his bloodline.

21 SEPTEMBER: Into the wood via the stile, and when I get to the beech a hallucination, a troop of wood blewits has appeared. Blewit is a corruption of 'blue hat', and the mushroom is true to its labelling, bluey-lilac, though the large cap browns with age.

Long ago, the mushroom was used to make blue dye in the clothing industry.

This most desirable of edible mushrooms has a delectable orange smell. I slip a dozen in the pocket of my coat.

22 SEPTEMBER: Blackbirds on the elderberries like plague, meaning the fruit is ripe, perfectly so. With their black brightness, the elder's berries make for a million mouse eyes.

I take a dozen carrier bags into the wood and load them with berries, stripping the lower tracts, leaving the upper ones for the birds, especially the addicted thrush family.

The berries are replete with Vitamin A (600 IU per 100mg), Vitamin C (36mg per 100mg) and antioxidants. So, well worth the picking for a cordial-cum-tonic that will see you through a sniffling winter (recipe: 600ml water, 225g honey, 25 elderberry heads, boil together, stirring all the while, leave overnight before bottling and consume within three months), or for a liqueur or a rich, port-like wine.

John Evelyn declared that an extract of the berries was a 'catholicon against all infirmities whatever'. The old diarist got it right by four hundred years; recent research has documented the berries' inhibitory effect on eleven strains of influenza as well as their capacity to stimulate cytokine production. Cytokines are cell messengers of the immune system, enhancing the organism's reaction to an infection. They also contain more antioxidants than most other small berries; one elderberry anthocyanin was found to be an effective inhibitor of human tumour cells *in vitro*.

Once more with meaning: in the health of woods lies our own health.

In historical times, the elderberry, the 'Englishman's grape', was grown commercially in elder orchards. Such is the versatility of the berry, it makes a delicious ketchup too, by simmering elderberries in equal parts cider and wine vinegar (just enough to cover the fruit in the saucepan), along with thyme, bay leaf, fennel and garlic salt. Bottle with a few peppercorns.

23 SEPTEMBER: At the edge of the wood, where the wind has its exits, the wind-music is bass toned in the trees, a soft susurration down in the grass; the lianas of honeysuckle, meanwhile, are plucked like strings.

24 SEPTEMBER: Sunny but windy; white butterflies; a lone damselfly; in the treetops greater spotted woodpeckers *chack*, the sound of two wooden boards clapping flat against each other. The woodpeckers are claiming winter territory.

25 SEPTEMBER: Rain, then sun, the pattern of our days. A few house martins hanging on.

28 SEPTEMBER: Pigeons roosting in the oaks, the ash and the top of the larch; they like an easy route to the freedom of the sky.

30 SEPTEMBER: Cannot sleep, so drive to Cockshutt and go for a walk in the wood. The pigs also restless. Under a fleeting moon and wild, smoky clouds, I see three hedgehogs trotting around the spotlighted glade, sniffing for slugs and worms, building up their reserves for winter.

OCTOBER

The Fruits of Autumn

Sallow leaves fall at the rate of three per second – the colour of rosehips – ivy flowers, and wasps – a spider's web – collecting crab apples for the pigs – service trees – autumn leaves: the moment of colour change escalation – the language of rain – fallow deer barking – the spruce lament their needles – why do trees lose their leaves? – dog fox barking – the storm

1 OCTOBER: At 4.41pm exactly: sun silvers the sallow, whose leaves drop at a rate of three a second. A late unseasonal burst of birdsong. With the canopy already part defoliated, there is less absorption, and the notes ping around the entire edifice of the wood.

3 OCTOBER: Rosehips turning Christmas-tangerine orange; hawthorn berries gone red, although as BB wrote about the haw: 'It is not a scarlet, but a rich crimson carmine, such as one sees in old brocade and velvet.'

The hazel tree: first to burst into leaf, first to burst into flame.

4 OCTOBER: Mmm. The soapy smell of ivy flowers. How to describe their appearance? A 3D model of a chemical molecule, but uniformly soft-sage-hued. The wasps, in their hordes, are on them, possessively, morning to night.

Shadows steal in; the wind blows a dirge through the stands of spruce.

The squirrel, shot with the .410, falls to earth with the thump of a boxer's glove.

6 OCTOBER: Early morning, and the brambles shining with spiders' webs.

Spiders are from the class of animals known as arachnids, named for the mythological maiden Arachne, who challenged Athena to a weaving contest and in punishment for her impudence was transformed into a spider. The creature shares with man and certain species of caddis fly the distinction of being able to set traps. The spider's snare is constructed of silk, produced from spinnerets at the rear of its abdomen. The silk is one five-thousandth of an inch thick; thickness for thickness, it is stronger than steel.

I'm sitting at the bottom of the Tall Oak, with the .410, waiting for squirrels. To save the birds, another two or three grey squirrels must die.

A spider starts her web on the bramble next to me; it takes her fifty-two minutes to weave her magic.

8 OCTOBER: A surprise: a new rabbit burrow, under a dingle-side ash, fresh earth spewed everywhere; such fine earth, no clods, the purest tilth.

Collecting crab apples for pigs, fir cones as fire lighters. The evening air, a foxy brown glow.

10 OCTOBER: The rams have prune-wrinkled foreheads. In trees the sap is descending: in the woodland animals the testosterone is rising.

12 OCTOBER: To my disappointment but not my surprise, there are no fruits on the service trees. They are too immature, I suspect.

Once, when I was young and walking around the village with my grandmother, we called in at Mrs Cole's cottage. She had a service tree in the copse behind her house, and asked Grandma and me whether we would like to try one of the fruits.

Mrs Cole retrieved a wooden box from her cellar, where she had put the service fruits to be 'bletted', stored until verging on rotten. I had my service fruit – a tiny brown pear – mashed on top of vanilla ice cream, and it tasted dangerously like adult sherry.

The service tree is also known as lizzory, checker, chequer. The country residence of the British Prime Minister, Chequers, is named after the service tree which grows in the grounds.

14 October: Under sun and cyanic sky Cockshutt burns.

16 October: Colour change escalation, for the worse; most trees have lost vibrant hues in favour of a muddy brown mixed by a toddler let loose on a paintbox of watercolours. Albert Camus asserted that 'Autumn is a second spring when every leaf is a flower.' Not here, not now.

And the sky is slate. You can forget the poets' mellow fruitfulness, milky-white mists, fire-red trees. On the Welsh borders, the rainfall is an inch higher now than in September.

Watch the willow leaves fall on to the pond. Each snowflake is unique. And the descent of every leaf is special, its death dance. We should be there to see their last glory.

Slopping through Lower Paddock, the air is thick with the sort of wet that frizzes human hair, and clams to a sheep's fleece to give it diamanté shine. Mutton dressed as princesses.

I put my face to the overcast sky to read the language of rain. It is mizzle, a meteorological signifier that drizzle, and worse, will follow.

I have my wife's black Labrador, Bluebell, with me.

Strange to say, a dog taught to retrieve makes a useful sheepdog (it can be sent about to block escape routes) and I drive the 120 sheep easily into the pens under the alders, on which the leaves are still clinging to life. Like the oak, the alder comes into leaf late, and defoliates late.

It is all change down on the farm in October. The wind has blown away the last straggling hirondelles; the self-same wind from the north brought the first redwings on its back edge. Thus the scene is replenished with birds.

Some of our sheep are also migrating. October is traditionally a month of sheep sales, so I spend the morning sifting and sorting the ewes and the lambs.

This little sheepy will go to Tenbury Wells market, this little sheepy will stay at home. Almost all the money in sheep farming is made in the months of September and October.

At eleven there is disco-shimmering drizzle; by twelve it is raining properly. The leaves of the alders are no shelter at all, but at least the lanolin in the sheep's coats waxes the front of my Barbour.

The rain does nothing to dampen the ardour of the five tups, who spend their hours in captivity trying to climb the galvanized hurdles to get at the ewes. The ewes are on heat, their hormones triggered by the decline in light. Under the low canopy of the alder the greasy musk of the aroused rams smothers like a foul pillow. Across in the dingle, the sheep's wild cousin, a fallow deer, is barking a come-on. Even rain dripping off alder fails to drown out his Duracell croaking.

I am finishing up dagging the sheep (trimming, with shears, the wool round their rear ends, which can get maggoty when impacted with faeces; it's all glamour down on the farm) when Bluebell, rootling about in the year-weary nettles, discovers the hedgehog. Mrs Tiggy-Winkle is unhurt by the encounter; the trademark prickly coat, which contains about five thousand spines, is effective against a Labrador's soft mouth, and softer brain.

After about fifteen minutes, Mrs Tiggy-Winkle unfurls. Deciding to risk the publicity of daylight, she ploughs through the soaking grass up to the base of the old brick barn, where slugs are swarming over windfall apples from a single domestic tree. She allows me the dubious privilege of standing close by. Her table manners are dreadful; selecting a large orange-trimmed slug, she bites at it, rolls it about, before finally gobbling it down. It is fascinating, but it is unnerving too. Nature can be very un-cute.

If Mrs Tiggy-Winkle was barrel-fat and ready for hibernation before, she is even better prepared after eating the slug. She waddles off towards the wood. Of course, if you think about it, hibernation is a type of migration. Mrs Tiggy-Winkle is off to the Land of Nod.

*

A kestrel flies up and sits on the telegraph wire, next to a wagtail. The young fox is lying under the brambles; he bolts out, seizes a rabbit. Surprise and speed. The fox can reach speeds of 40mph.

For a woodland fox, life will never be better. The rabbit population is at its yearly peak. There are currently around sixty rabbits in Cockshutt.

Autumn

I saw old Autumn in the misty morn
Stand shadowless like Silence, listening
To silence, for no lonely bird would sing
Into his hollow ear from woods forlorn,
Nor lowly hedge nor solitary thorn; –
Shaking his languid locks all dewy bright
With tangled gossamer that fell by night,
Pearling his coronet of golden corn.

Where are the songs of Summer? – With the sun,
Oping the dusky eyelids of the south,
Till shade and silence waken up as one,
And Morning sings with a warm odorous mouth.

Where are the merry birds? – Away, away,
On panting wings through the inclement skies,

Lest owls should prey
Undazzled at noonday,
And tear with horny beak their lustrous eyes.

Where are the blooms of Summer? – In the west,
Blushing their last to the last sunny hours,
When the mild Eve by sudden Night is prest
Like tearful Proserpine, snatch'd from her flow'rs
To a most gloomy breast.

Where is the pride of Summer, – the green prime, –
The many, many leaves all twinkling? – Three
On the moss'd elm; three on the naked lime
Trembling, – and one upon the old oak-tree!
Where is the Dryad's immortality? –
Gone into mournful cypress and dark yew,
Or wearing the long gloomy Winter through
In the smooth holly's green eternity.

The squirrel gloats on his accomplish'd hoard,
The ants have brimm'd their garners with ripe grain,
And honey bees have stored
The sweets of Summer in their luscious cells;
The swallows all have wing'd across the main;
But here the Autumn melancholy dwells,
And sighs her tearful spells
Amongst the sunless shadows of the plain.

Alone, alone,
Upon a mossy stone,
She sits and reckons up the dead and gone
With the last leaves for a love-rosary,
Whilst all the wither'd world looks drearily,
Like a dim picture of the drownèd past
In the hush'd mind's mysterious far away,
Doubtful what ghostly thing will steal the last
Into that distance, gray upon the gray.

O go and sit with her, and be o'ershaded
Under the languid downfall of her hair:
She wears a coronal of flowers faded
Upon her forehead, and a face of care; –
There is enough of wither'd everywhere
To make her bower, – and enough of gloom;
There is enough of sadness to invite,
If only for the rose that died, whose doom
Is Beauty's, – she that with the living bloom
Of conscious cheeks most beautifies the light:
There is enough of sorrowing, and quite
Enough of bitter fruits the earth doth bear, –
Enough of chilly droppings for her bowl;
Enough of fear and shadowy despair,
To frame her cloudy prison for the soul!

Thomas Hood

17 OCTOBER: Sibelius maintained that trees talked to him. Standing next to the spruce in the wind, I believe him. They lament their needles, which catch the wind, and whisper enviously of the deciduous trees, more naked by the day and so less affected by blow.

Why do trees lose leaves? Broadleaved trees get so little light in winter that it is not worth them expending their energy on keeping dressed until spring. And as every sailor knows, one reduces sail in a wind. In trees sap retreats and chlorophyll in the leaves breaks down to expose yellow and orange pigments hitherto hidden.

Sycamore seeds, which are winged, helicopter for yards away from their parents.

18 OCTOBER: 4pm. Air clear and dry; the robin's song a bell tink, perfect tone, clarity in the cold.

Measure a spider's web, and it is eighteen inches across.

Pigeons fly from wood to wood to land on the oak, and cover them, like lice. A flock of a hundred, maybe? The Victorian naturalist W. H. Hudson, in *Nature in Downland*, saw a wood pigeon flock which 'could not have numbered less than two to three thousand birds'. Even our pests are diminished in the new farmland. The combined sound of the pigeons' wings is sheet metal, shaken.

A hen pheasant slinks on seeing me, but I am no threat, I am lost in the High Church incense of autumn woodland. The dark notes of autumn – earth, wood, decay, fermentation – are enough to send one into a reverie.

The smell of decay is beautiful and deadly. The year is dying.

Who would be absent from England in October?

19 OCTOBER: Ash trees die pale and greeny, sickly Victorian heroines. I guess at least half the acorns have fallen (a modest crop this year), wizened brown, but yellow where they were fixed in their cups. In good years the acorn crop is prodigious; they are planted by mammals and birds – one reason for their success as a species. Jays secrete acorns in hawthorn, safe from grazing animals. The hawthorn is the mother of the oak, the most widely distributed of forest trees.

Dog fox barking in Cockshutt at 6.30pm, and a tawny over my head, unseen in the canopy, sub-text.

Stars are white sparks in the sky. Sitting in my chair I watch a hedgehog, with its snail-wet nose, feasting on a slug.

20 OCTOBER: The corroding tint of russet on the wood top sinks deeper by the day, a descending stain.

A few leaves still on the sallow wands, quite like pennants on a cavalryman's lance, shredded in the wars of the winds.

Three robins singing for territory; a nuthatch hammers a hole in an acorn, which it has wedged in a crack of the Wishbone Oak.

I think I see a kingfisher on the pool, more by what it does than by colour or shape; it flies off a perch on the far bank, snatches something from the water, returns to the perch, then zips off. But it is dark under the trees, not because of leaves but because the alders, willows and hazel lean menacingly over the water.

Sitting in my chair: looking left into spruce, commercial Norway, right into deciduous, old Eng-la-land. A squirrel: *chak-chak-chak-chee*. The squirrel runs through the high branches, a grey streak of lightning.

A small black beetle scuttles about. (Saxon *bitela*, meaning bitter, refers to the creature's acrid taste.) Resolve to go on a beetle hunt.

21 OCTOBER: The storm came in the night. The forecasters gave it a girl-next-door name, Katie or some such, but really she was Maleficent. This morning, standing beside the wet beech trees at the bottom end of Cockshutt, the scene is a magnified pick-up-sticks. The woodland floor is a chaos of crashed branches and

downed trunks. Among the fallen is the beech by the stile, the one I touch every day as I pass.

Touched. Past tense.

I look up through the vertiginous gap she – yes, to my mind she was always the empress of the wood – left. There is a hole in the roof of the wood. There is a hole in my day.

23 OCTOBER: Showers of leaves. Since there is no one about I run around like a kid, trying to catch the leaves in mid air. I did the same, in simpler times, when the children were younger.

24 OCTOBER: A small party of redwings arrive in the meadow next to Cockshutt. Their calls knock nails in the coffin of summer.

Pin-thin cries of goldcrests in the larch.

At night, checking the pigs with a torch: overhead the steady movement south of Arctic thrushes. Summer's bird migrants come for the meat; the winter migrants come, predominantly, for the fruit.

Finally I go on a beetle hunt in Cockshutt, leaving no log (gently) unturned, no hollow in a tree uninspected, and in twenty minutes find a violet click beetle (*Limoniscus violaceus*), a cardinal click beetle (*Ampedus nigerrimus*), another click beetle (*Elater ferrugineus*), a false click beetle (*Eucnemis capucina*), and a wood-boring weevil (*Dryophthorus corticalis*).

25 OCTOBER: About a third of the leaves have changed colour; almost at a stage where the leaf litter equals the amount of leaves left on the tree. To stand under the crab-apple trees is to be inside a golden temple.

Pitter-patter of falling leaves; bomb-bang of falling acorns.

Sitting in my chair, an electric jolt – three fly agaric mushrooms have grown in the birch. Topple one with my boot; the underneath is seductive, luxury cream; the top is seductive, luxury scarlet, thick like icing or sugar glaze. Everything about a fly agaric says 'Eat me.'

Viking warriors did eat small amounts of fly agaric as preparation before going into battle. The mushrooms contain a mescaline compound, which affects the nervous system in the manner of LSD. Consumption of the mushroom's flesh would cause a loss of inhibitions and fear, spurring the Norsemen to feats of bravery, born of delusion. The Vikings took half of England because they were berserk drug-takers.

26 OCTOBER: The first fieldfare; a bewildered tourist, looking around.

The leaves of the service trees flutter, flicker, flame, in pseudo fire.

Tree leaves lie on the pool: false lilies.

All through Cockshutt Wood the ash trees are fading elegantly, while the hawthorns proffer bright berries to thrushes.

The last days have been so cool that evaporation from the ground is slow, if at all, and fog follows mist, and mist follows fog.

Oh, the melancholy of autumn woodland.

28 OCTOBER: The day everything changed, the clamp of winter; a change of posture and mood among animals, trees and plants alike. To look out of the house windows is to feel cold.

The *clip-clop* of Saturday horses on the wet lane.

About half the leaves in Cockshutt are now down, compared to only a third three days ago.

But it is cold now, and cold loosens and snaps off the leaves. An abscission layer forms on the leaf stalk, weakening and finally severing its attachment to the twig from which, in spring, it grew. Fallen leaves are the beginning of future humus.

Back to collecting kindling. Or how humans mark the coming of the cold. It is almost the full turn of the natural year since I began this diary.

29 OCTOBER: Overcast night. Under leaf canopy, absolute blackout.

On such a night, when the eyes are not enough, Old Brown locates his prey by listening to movement. The ears of the owl are marvellous.

In owls, the ears are set asymmetrically in their skulls, with one ear as much as fifteen degrees further up the skull than the other, and sometimes of larger size. Asymmetrical positioning of the outer ears means that each ear receives sound at a slightly different volume and angle, which allows the owl to pinpoint where the sound comes from. There is also a flap of skin in front of the ear which the bird can control to catch sound, in the way that old gentlemen cup their hand to their ear. Indeed, the entire facial disc of the owl acts as an amplifier.

Night is no friend to prey animals when an owl is about. The wood (to my ears) has slept into silence. Old Brown, however, can hear the turn of a leaf . . . and the scamper of a rabbit on the grass of the glade.

The scream comes from only yards away, a scream that terrifies the night. I and all the little animals in the wood stop – still as statues in musical-chairs – and hold our breath.

It is a distinctive, high-pitched wail, well known to us wood folk.

The Lord of the Night has killed a rabbit.

30 OCTOBER: The bats have quit the attic to roost in the rotten ash in the dingle, the one peeling to reveal its interior. Once, in a spirit of Gilbert Whitean nature observation, I climbed up the tree and poked a torch in the hollow where the hibernating colony hang. I expected the pipistrelles to be inert; instead, they shuffled en masse and raised their heads, a sort of living wall with a hundred faces. It was fascinating, but it was unnerving too.

I take Rupert the Border Terrier for a walk in the wood.

Autumn glow tinged by a cold blue, which has scourged the sky of cloud; the fun of scuffing through leaves.

The dwindling metabolism of the toad.

Ash leaves do not fall singly but in brackets of six to eight. Crab-tree leaves fall in a shower of golden pennies; the money tree. Sweet chestnuts lie on the ground in pocketfuls.

Native to western Asia, sweet chestnuts were introduced to southern Europe by the Ancient Greeks, and brought across the Channel by the Romans. Although the tree has naturalized in Britain, the two or three nuts or kernels contained in the spiky British burr are smaller than their continental counterparts. The sweet chestnut, despite its name, is not related to the inedible horse chestnut, the 'conker tree'. Due to its mass of spines, country people called the sweet chestnut the 'hedgehog'.

The nuts can be eaten raw, but this is a waste. Raw they are tough and unremarkable; cooked they are glorious.

More than a third of sweet chestnut kernel is made up of carbohydrate, making it a staple source of starch in some regions of southern Europe. The Italians – among a hundred other uses for the sweet chestnut – boil the kernels as a vegetable, as well as grinding them into flour for polenta, bread and cakes. To make chestnut flour, roast the chestnuts, then grind: the resultant flour is sweet and yellow but does not rise well. The French, of course, have candied the chestnut into *marron glacé*. In Britain, the chestnut has tended to be viewed as a sa-voury, the stuff of stuffing and that Edwardian country house classic, chestnut soup. And, of course, the chest-nut roasting on an open fire is as redolent of Christmas as cathedral choirs singing 'Good King Wenceslas', huntsmen on horses and the Queen's Speech. If you don't have an open fire, an oven will do. Simply lay on a baking tray and roast at 200°C. But whether you are roasting on a fire or in an oven (or indeed boiling), you must make a small X-shaped slit in the shell before exposing to heat, otherwise they explode spectacularly in a shower of sweet chestnut shrapnel. Boiling makes the bitterish inner skin easy to peel off.

> *The gathering of these nuts is a great business in the south. The woods become places of pilgrimage, not so much for the country folk, who have an astonishing disregard for the fruits of the earth lying at their back door, as for the towns-folk. Out they come, by bus and car and bike, principally men, very earnest, and principally on Sunday mornings. It is a piece of yearly ritual. They raid the woods like human squirrels, spending hours kneeling or stooping or even sitting under the canopy of leaves already much thinned by rain and frost, foraging among the blanket of fallen leaf and husk, filling cans and sacks with silk-soft nuts, staggering out at last under the weight of their pot-bellied sacks, still looking very earnest but, somehow, satisfied.*
>
> John Stewart Collis

Chestnut purée

Based on the recipe by Elizabeth Craig in *Cookery Illustrated and Household Management*, 1936

900g chestnuts	75ml milk
50g butter	salt and pepper
vegetable stock	caster sugar

Cut the tops off the chestnuts and roast them in the oven for 20 minutes. Remove the outer and inner skins, and put the chestnuts into a stew-pan with half of the butter and enough stock to cover them. Lay greaseproof paper over the top, put on the lid, and simmer for 45 minutes or so, until the chestnuts are tender. The chestnuts should absorb all the stock in the cooking.

When cooked, rub all through a fine sieve. Thoroughly mix this purée with the remaining butter and the milk. (You may not need all the milk, depending on how much stock has been absorbed.) Season with pepper, a very little salt, and a pinch of caster sugar. Re-warm and serve with venison loin.

Or, minus the pepper, fill a crêpe, or load on top of vanilla ice cream.

Chestnut soup

My favourite way of utilizing sweet chestnuts.

Serves 4

675g chestnuts	1 sprig rosemary
1 onion, finely chopped	chicken or game stock
1 carrot, finely chopped	150ml single cream
30g butter	parsley, finely chopped

Cut crosses in the ends of the chestnuts, place in a pan with enough water to cover and boil for 2–3 minutes. Remove from heat, and when the chestnuts are cool enough to handle, peel, scrape off the papery inner skin, and put them to one side.

Sweat the onion and carrot in the butter until tender. Add the chestnuts and rosemary and continue sweating over a low–medium heat for 5 minutes. Pour in the stock, then simmer for 20–30 minutes. Liquidize the soup, then strain into a clean saucepan and add the cream. Bring up to almost boiling and season to taste. Serve with a scattering of chopped parsley on top.

31 OCTOBER: The clocks go back: next week at 6pm the Earth's revolution will be an hour further onwards than at 6pm the week before. The cars on the lane coming home will have their lights on, past cottages lit with television screens. The year has turned – it cannot be gainsaid. The year has turned.

And in Cockshutt the trees know it.

I harvest my thoughts in this end of time. It is the nature of the oak to be still; it is the nature of the kestrel to wander with the wind.

A leaf falls slowly, as if lowered on a thread. A moth on a branch has his wings half open, like a little beige coat over his shoulders.

The animals have plundered the hazelnuts, ruins and remnants of the latter all across the woodland floor. I admire best the nuts which have been turned round and round as the mouse gnawed its hole.

Bright masses of haws in the sun; great brown cones fallen from the spruce firs; green and rubine berries of white bryony hanging thickly on bines from which the leaves have withered.

Owls: An Epitaph

What is that? . . . Nothing;
The leaves must fall, and falling, rustle;
That is all:
They are dead
As they fall,
Dead at the foot of the tree;
All that can be is said.
What is it? . . . Nothing.

What is that? . . . Nothing;
A wild thing hurt in the night,
And it cries
In its dread,
Till it lies
Dead at the foot of the tree;
All that can be is said.
What is it? . . . Nothing.

What is that? . . . Ah!
A marching slow of unseen feet,
That is all:
But a bier, spread
With a pall,
Is now at the foot of the tree;
All that could be is said.
Is it . . . what? . . . Nothing.

Edward Elgar, 1907

NOVEMBER

Out of the Woods

*Yellow brain fungus – an owl pellet – the animals race
against time – crab apples – acorn coffee – woodcock –
Richard Jefferies on how to tread lightly, like an amateur
poacher – planting acorns – my father – fieldfares – my
last walk in the wood*

2 NOVEMBER: 3pm. Refreshingly chill, but blinding
Gestapo sun in the west. A touch of mist from breath;
rams literally panting.

The smells of autumn: wet cardboard but with kiwi
fruit undertone.

Beech leaves on the ground; copper pennies, another money tree. Oak leaves – clotty brown pancakes
underfoot, but with tobacco smell (and an instant
Proustian trigger to remembrance of my pipe-smoking
father, RN in World War Two).

Elder leaf: a wet, limp handshake of a leaf.

On a hazel stump, yellow brain fungus (the name is
its description), and crows wearing their widow's weeds
in the larch top.

The gean is the first tree to lose all its leaves. At the
foot of the gean an owl pellet, a felt cabinet of skeletons
and curiosities. Gingerly, with sticks, I prise apart the
pellet, with its gut-bleached bones. The tawny's diet is
dominated by field voles, bank voles, wood mice, large
beetles, birds and moles. (With a body weight in excess
of 430g the adult female tawny is big enough to take

273

young rabbits.) Pellets pack surprises. In this pellet I find the skull of a newt. How Macbeth's witches would have loved that.

5 NOVEMBER: The owl-light.

On the south horizon, May Hill has already drowned. The visible world is shrinking to our fields and wood. And God, is it freezing; I am cripple-hunched over the tractor steering wheel as I drive along the pig field. The Ferguson is cabless, but happened to be handy and working, unlike the Land Rover; hitched on the tractor's rear is the transport box, with a sack of sow breeder nuts and a standing Labrador trying to keep her balance. The metal of the box is ice-tacky, too cold to sit.

The kestrel, backlit by the high half-moon, quarters the maize stubble on her last hunt of the day. Something catches her attention; she anchors, drops, then flickers away triumphant.

Migrant birds bring change, but the kestrel is an emblem of eternality, a reminder of what in the countryside is constant.

6 NOVEMBER: Explore the coppiced holly trees, their leaves smooth and flash. Very druidy; a dark place in a light wood. The birds and the animals are in a race

against time: fighting for the berries on holly trees, the hawthorns, the red hips of the briar, the blackberries, the acorns, the beech.

7 NOVEMBER: 4.50pm, cold clear sky, an ice star in the west, over Garway.

I walk the 'wrong way' around the top of Cockshutt, pushing through the briars, hazels, bines of the eastern edge up towards the foxes; I see the wood anew, and notice a diagonal rabbit track leading from a fresh warren.

Three jets in the sky; flaming red arrows.

On the pool a moorhen dives under water, with the professional speed of a U-boat.

10 NOVEMBER: From somewhere ahead Old Brown calls *kerwick*. I have a watch, but I do not need it now. He has clocked the coming of night.

Trees claw at the sky for breath.

12 NOVEMBER: As much as I love the pigs, giving them absolutely all the crab-apple windfalls is a mite too generous, so a carrier bag's worth goes home with me.

*

Sour grabs. Scarb. Scrab. Bittersgall. Gribble. Scrogg. They could be names for hobgoblins, but actually they are names for the crab apple.

The wood, which is very hard, was used in the past for wood carving, sauce. And punishing wives:

> The crab of the wood is sauce
> Very good for the crab of the sea,
> But the wood of the crab is sauce
> For a drab who will not her husband obey.

Malus sylvestris is native to Europe, and is the ancestor of the cultivated apple.

The wild crab differs from cultivated and feral apples ('wildings') in being thorny, the smallness of its fruit (two centimetres or so) and having hairless leaves and flower stalks.

The country names for crabs serve as a warning to the unwary biter of this wild fruit: it is mouth-puckeringly acid. In some localities, the cider made from crabs is known as 'pig-squeal cider', because the drinker inadvertently shrieks like a porcine on tasting the brew. The astringent fruit gave rise to the English word crabby, meaning bad-tempered. In medieval times, crab apples were used for the sour sauce verjuice, just as the modern cook uses lemon.

You will need to sweeten your crabs in the cooking. Crab-apple jelly is the best-known way to use the fruits, not least because they are pectin-rich and help setting. But there are many other things you can do with crab apples; when Julius Caesar invaded Britain in 55 BC he found the Celtic natives fermenting the juice of *Malus sylvestris*. The Romans titled the drink *Sicera*. Cider.

Crab-apple jelly

Ideal on toast for breakfast, ideal as an accompaniment to roast meat. This recipe makes 2–4kg of jelly.

around 4kg crab apples   1 lemon
around 2kg caster sugar

Wash and quarter the crab apples, but leave in the cores because they contain heaps of the pectin necessary to set the jelly. Put the apples in a large heavy pan or cauldron with just enough water to cover them. Bring to the boil and simmer until the fruit is pulpy and soft (about 25–30 minutes).

You now need to pour the pulp into a jelly bag – or double layers of muslin – and let it drip into a pan overnight. (Tying the corners of the bag to an upside-down stool with a bowl underneath to catch the apple drips is the ideal arrangement.) If you hurry the process by squeezing the bag this will make the jelly cloudy; if you are unworried by aesthetics, squeeze every last drop out.

Next morning, measure the apple juice and add sugar in the ratio of 500g sugar to 1 litre juice. Tip into a saucepan, add the juice of the lemon and bring to the boil, stirring with a wooden spoon to dissolve the sugar. Remove any surface scum. Boil hard for ten minutes, then test for the setting point with a sugar thermometer: this is 105°C. Alternatively, have a fridge-chilled pudding spoon on standby and put a small amount of the jelly on the back of the spoon. If it solidifies it is set. If it is still liquid, boil some more, then repeat the test.

When the setting point is reached, remove from the heat and ladle the jelly into warm sterilized jars. Cover with a waxed paper disc, tightly seal with a lid, and store in a cool dark place. The jelly should keep for a year. A more savoury jelly can be obtained by adding herbs, such as sage and rosemary.

The recipe for crab-apple jelly can be used as the model for a hedgerow jelly using rosehips, haws, sloes, bullaces, rowanberries, elderberries or blackberries in any proportions you can forage, as long as the mix includes 50 per cent of the high-pectin crab apple.

13 NOVEMBER: Leaf-stripper November wind has uncovered a blackbird's nest in a hawthorn bush. The nest sits like an accusation; the inside is full of mice-dined haw kernels, and leaves so long dead they are skeleton. Live moss, glossy and green, has begun its sybaritic hegemony, at once gorgeous but unlovely. (We do not like moss.)

How did I not see the nest last year when it must have been full and fecund with eggs, and chicks?

14 November: A murmuration of jackdaws, ninety or more. Only the oak leaves show; the larch is golden.

In my chair: a companionable wood mouse eats an acorn beside me. If the wood mouse did not gnaw, its teeth would grow to such an extent they would be tusks and it could not eat.

For humans, acorns, thick with tannins, are inedible raw, though in times and places of austerity, such as Germany in the Second World War, acorns have been roasted and ground as an ersatz coffee.

Acorn coffee

Take the acorns out of their cups, add to a saucepan of boiling water and roil for 10 minutes. This is to soften the shells. Drain, let cool and peel. Leave to dry for a day or two. A windowsill or airing cupboard is ideal. Then roast on the middle shelf of the oven, 120°C/Gas Mark ½ for about 15 minutes. Grind. Use the acorn coffee as you would ground coffee; put into the cafetière or percolator at the rate of 2 teaspoons per cup. The acorn coffee can be kept in airtight jars or tins.

Acorn coffee tastes nothing like coffee. It is, however, a perfectly pleasant beverage, something like the health food shop favourite, Barley Cup.

The acorn does seem to be a nut most valued in

times of privation. In the second century AD the Roman doctor Galen recorded poor country folk making flour from acorns. Again, the acorns need to be boiled and de-shelled. Then place the acorns in a muslin bag and immerse in water for two weeks or more, taking care to change the water twice a week. To dry them, place in a sunny spot (inside, out of the reach of birds) or in the oven at low heat. When the acorns are completely dehydrated, grind and place the flour in a paper bag. Like the flour of sweet chestnut, it will not last long in the cupboard before going mouldy.

Usually, oaks do not produce acorns (which, botanically, are the fruit of the tree) until they are forty, and trees are often biennial.

15 NOVEMBER: Dusk is already seeping through the trees like silt when I go to look for the deer. Up into the wood, past the alder, the beech, into the stand of spruce. Here the accumulated century of pine needles has made the ground springy, yielding. Silent.

Wind, undiluted from the Urals, enfilades Cockshutt, although this evening I count it my robber friend; it steals my smell away. Out in the wintry towers of the oak are the five fallow deer, hungry enough to eat the iron rations of bramble leaves. I get to within fifty yards of them. Forty yards. Thirty, before they exit.

16 November: Torrential rain; leaves on the floor, flattened. Patterned lino.

Time and light are failing. I follow the faint ink-line of the path as it squiggles between the dulled obstacles of beech, sweet chestnut, hornbeam. A single stick, hidden under the wet vinyl of leaves, snaps; the cannon *boom* around the vast empty chamber of the wood sends pigeons clattering through the tops of the trees.

The naked trees. Every last leaf was stripped off in the storm. In twenty years I cannot remember such a violent undressing. (It was a north, Viking gale.) In a day the wood was transformed beyond recognition.

Walking quicker now. Dog-trot. To my left, glimpsed between passing trunks, a finger-smear of dying sun.

The more the blindness, the greater the sense of smell. Ah, the full autumn Bisto bouquet, the sweet smell of rot, comes powering to the nose: mouldering leaves, decaying mushrooms, rusting earth.

I'm just skirting the dingle when the woodcock explodes from under my foot. An avian IED. I shout out unmanfully in the silence. Luckily, in a wood on the far edge of Herefordshire there is no one to hear me scream. Unless the Jew's ears on the elder heard my cry, and translated it for the trees. Is the elder the listening tree?

Softly, softly now. Richard Jefferies, author of *The Amateur Poacher*, provided the masterclass on how to tread softly: 'The way to walk noiselessly is to feel with the foot before letting your weight press on it; then the dead stick or fallen hemlock is discovered and avoided. A dead stick cracks; the dry hollow hemlock gives a splintering sound when crushed.'

I can hear the cattle moving; the crackle of branches under hooves, the slow drumbeat of massive beasts in motion.

There they are, out in the towers of oak, four cows, prehistoric shapes plodding up to the top of the wood.

Cattle come from the old time, and are no strangers to death amid the trees.

And I am the killer in the wood. The little tricks: do up coat and collar to break face; hunch the body to look less like a man; scope floor to treetops. A few, shallow, regularizing breaths . . . I press off the safety catch of the shotgun, and slip into the noise shadow of the cattle.

The cattle shuffle under the extended, flamboyant ballerina-arm of an oak, on which rests the silhouette of a cock pheasant . . .

The flash-blast of the shotgun rips the wood apart. The cattle trumpet alarms. The tawny owl cries out.

The pheasant plunges head first, tail streaming behind. A black comet falling to Earth.

In the poker game of life and death we all have our tells. The pheasant had roosted on the same branch for a month, each night dropping his white guano on to the ground.

As I pick up the pheasant, a gap comes in the clouds.

The North Star shines brightly, more brightly than usual.

This is *Brendon Chase* for grown-ups. And why not? So, I say again: I manage the wood for wildlife; should not wildlife in return provide me with a meal?

17 NOVEMBER: The wind gets up; the close-planted larch brush and rattle each other. There is a certainty of a storm in November, like there is of a shower in April.

The autumn wood: vitality has gone, and in the trees we witness our own ageing.

18 NOVEMBER: Since there are no jays, nature's master oak planters, in this wood I do the planting. I, the apprentice, spend the afternoon with a spade, slitting the earth and popping in acorns, here, there, everywhere. Wrens sing all day from the understorey.

Somewhere, from over the hill, the caw of a rook rolls over the sodden land. Rooks are also apprentice planters of acorns. They will take and bury acorns, as much as a mile from the provender tree.

19 NOVEMBER: The hardest thing in watching nature – to be here, now. Not thinking about something else, as I am doing. (But I do notice that the oaks and hazel are still leafy, and the ash is already forming buds.)

So: we are moving house, going to the other, eastern side of Herefordshire, Ledbury way. Only twenty miles, but a big thing for us after twenty years of hill farming on the wild, west edge of the county. A new life beckons, that of growing hops; it's an old life too, the one I knew in childhood . . .

I am very proud of my late father who, aged seventeen, joined the Royal Navy to fight for his country, and for several years of my childhood raised me alone.

During packing, I find his pension policy with Hearts of Oak.

For more than a century, Hearts of Oak, which was founded in the Bird in Hand pub in London's Covent Garden in 1842, operated as a pensions business. It is difficult to think of a more appropriate pension provider for an old seadog.

In a Wood

Pale beech and pine-tree blue,
Set in one clay,
Bough to bough cannot you
Bide out your day?

When the rains skim and skip,
Why mar sweet comradeship,
Blighting with poison-drip
Neighbourly spray?

Heart-halt and spirit-lame,
City-opprest,
Unto this wood I came
As to a nest;
Dreaming that sylvan peace
Offered the harrowed ease –
Nature a soft release
From men's unrest.

But, having entered in,
Great growths and small
Show them to men akin –
Combatants all!
Sycamore shoulders oak,
Bines the slim sapling yoke,
Ivy-spun halters choke
Elms stout and tall.

Touches from ash, O wych,
Sting you like scorn!
You, too, brave hollies, twitch
Sidelong from thorn.
Even the rank poplars bear

Illy a rival's air,
Cankering in black despair
If overborne.

Since, then, no grace I find
Taught me of trees,
Turn I back to my kind,
Worthy as these.
There at least smiles abound,
There discourse trills around,
There, now and then, are found
Life-loyalties.

Thomas Hardy

22 NOVEMBER: They came in the night. I heard them. *Chacka-chacka-chacka.* I have been waiting for them. In an early morning of squinting diamond-brightness I go to the wood to see them. The ground is gone to white and iron, so they are up in the hawthorns, pillaging, plundering, robbing.

Fieldfares. More than thirty of them.

The grey birds signal the arrival of winter as surely as plunging mercury.

Winter trails in their wake as they wing down from the north. Chaucer, in 'The Parlement of Foules', dubbed them the 'frosty feldfares', because their first coming so precisely coincides with cold. (Some locals, long in the artificial tooth, still call the fieldfare the snowbird.) They say timing is everything, and nature does synchronicity perfectly; the haws in the wood are lipstick-glossy with ripeness, and give themselves in shameless kisses to the birds.

By a mystery I cannot fathom, the fieldfares have coincided their migration with the availability of berries on bushes on a small, high farm in Herefordshire, a county unknown to most people, let alone avifauna. Perhaps, though, I am not meant to understand; perhaps I am meant only to wonder. The seventeenth-century Herefordshire poet Thomas Traherne believed man fell from a state of innocence because he turned from nature to intellect. There are days, days like these when I am looking at birds which have travelled a thousand and more unerring miles, when I cannot disagree with Traherne's *Centuries of Meditation*.

It is in the thrill of winter that you can feel the terrible, lovely contradiction of nature best. Beauty and barbarism always go together. Haw and thorn. This Canaletto-blue sky, this stark frost. Yin and yang.

There is no fear in the eye of the fieldfare. They are big thrushes with a flock mentality; the local sparrowhawk, attracted by the commotion that is fieldfares breakfasting, slips a low circuit around the wood. He is a twisting blade of badness. The fieldfares let rip a machine-gun pattering of *chacka-chacka-chacka!* The sparrowhawk, no fool, veers off for easier pickings, down the hill over the sheep fields towards where the mist lies in a pearl sea.

Within two hours the fieldfares have plundered to their fill. Ever restless, they suddenly lift into the air, and away. *Felde fare* is Anglo-Saxon for traveller over fields. I too must go travelling, on my farming round.

By 6pm ice is clamping the world, it is darkening fast, and I am still chopping wood for the log burner so we have heat in the cold.

Down in the valley, a fox yaps away, which is Reynard's equivalent of chatting on Tinder. In the dying of the year, the foxes are beginning their cycle of life. Then, as I'm stacking logs by the field gate the Cockshutt fox shoots past me and cat-arches on to a rabbit; the rabbit screams into the night. Haughty in his firestorm coat, the fox gives me a dismissive angel-eyed stare, then trots into the dark.

The rabbit is still wailing in the fox's jaw. Suddenly this remote world is very cold. Under the starlit sky, more fieldfares come over.

24 NOVEMBER: Deep into the wood now, past the sheeny boles of beech, the ground dry here. Crackle, snap of twigs under wellingtons; vandalism on this silent night, despite my cat-walk efforts. Weaving in and out of the trees, into shadow then out into the

moon's glare. Black and white. Ahead, the bob-flash of a running rabbit.

The woodcock whirr away, banking left, banking right through the moonlit beeches, flying on an instinct perfected down the centuries to escape falcons and men bearing guns. In their escape, the woodcock make a sound like ripping paper.

In the stark light of the moon I pull off a row of the Jew's ears from the elder, and stick them in the breast pockets of my boiler suit. The mushrooms will be fried with egg as tomorrow's breakfast.

The elder tree gives me a Judas kiss; I forget to duck, and get a whip from a branch across the face.

Fatty moonlight gleams on the pool.

29 NOVEMBER: During the early day, trees on the horizon line up, lungs on the shelf of a medical display.

The pool is frozen. Frost calls down the final lingering leaves.

I take my last walk through the wood. The dingle, as ever, is a degree lower in temperature than the rest of Cockshutt.

I touch the oaks, give the BFG redwood a friendly punch, blow a kiss to the moorhen . . .

It is the end of my tenancy of the wood. I may never see it again. How will the trees, the birds, the fox, the butterflies get along without me? I noted it all so carefully, day by day. Every tree, wood anemone, owlet, fox cub, beetle, mushroom and berry. It is not easy to say farewell.

I thought the trees and the birds belonged to me.
But now I realize that I belonged to them.

Sources

A Woodland Reading List

H. E. Bates, *Through the Woods*, 1936, reprinted 2010

BB (Denys Watkins-Pitchford), *Brendon Chase*, 1944, reprinted 2000

John Stewart Collis, *The Wood*, 1947, reprinted 2009

Roger Deakin, *Wildwood: A Journey Through Trees*, 2007

John Evelyn, *Sylva, or A Discourse of Forest-Trees and the Propagation of Timber*, 1662, new edn 2014

Richard Fortey, *The Wood for the Trees: The Long View of Nature from a Small Wood*, 2016

Geoffrey Grigson, *The Englishman's Flora*, 1975, reprinted 1996

Nick Groom, *The Seasons: An Elegy for the Passing of the Year*, 2013

Thomas Hardy, *The Woodlanders*, 1887

W. H. Hudson, *Hampshire Days*, 1903, reprinted 2016

Richard Jefferies, *Field and Hedgerow*, 1889

Thomas Pakenham, *Meetings with Remarkable Trees*, 1996

Oliver Rackham, *The History of the Countryside*, 1986, illus. edn 2003

Jeffrey Radley, 'Holly as a Winter Feed', *The Agricultural History Review*, Vol. 9, no. 2, IX, 1961

Eric Simms, *Woodland Birds*, 1971

Martin Spray, 'Holly as a Fodder in England', *The Agricultural History Review*, Vol. 29, no. 2, 1981

David Streeter and Rosamond Richardson-Gerson, *Discovering Hedgerows*, 1982

Edward Thomas, 'The Maiden's Wood', in *Rest and Unrest*, 1910

——, *The Woodland Life*, 1897

Mike Toms, *Owls*, 2014

Colin Tudge, *The Secret Life of Trees*, illus. edn 2005

Wood Music: A Playlist

Foals, 'Birch Tree', 2015

Arnold Bax, November Woods, 1917

Beach Boys, 'A Day in the Life of a Tree', 1971

The Beatles, 'Norwegian Wood', 1965

William Boyce and David Garrick, 'Heart of Oak', 1760

George Butterworth, The Banks of Green Willow, 1913

——, 'Loveliest of Trees', from 'A Shropshire Lad', 1911

Editors, 'I Want a Forest', 2009

Edward Elgar, String Quartet in E minor, Op. 83, 1919

——, Quintet in A minor, Op., 84, 1918

——, Cello Concerto in E minor, Op. 85, 1919

——, Owls: An Epitaph, Op. 27, 1907

Keane, 'Somewhere Only We Know', 2004

Lindisfarne, *Dingly Dell*, 1972

Oasis, 'Songbird', 2002

Pink Floyd, 'Careful with That Axe, Eugene', 1969

Camille Saint-Saëns, *Le Coucou au Fond des Bois* ('The Cuckoo in the Depths of the Wood'), 1886

Pablo Casals, *El Cant dels Ocells* ('Song of the Birds'), 1961

Antonín Dvořák, *Waldesruhe* ('Silent Woods') for cello and orchestra, Op. 68, no. 5, 1894

Edvard Grieg, Lyric Pieces, OP. 43, no. 4, 'Little Bird', 1886

Franz Liszt, Legende S.175 no. 1, St Francis of Assisi preaching to the birds, 1863

Monty Python, 'The Lumberjack Song', 1975

Van Morrison, 'Redwood Tree', 1972

Wolfgang Amadeus Mozart, *Der Vogelfänger bin ich ja* ('The Bird-catcher, that's me'), from *Die Zauberflöte* (*The Magic Flute*), 1791

George Perlman, 'A Birdling Sings', from 'Ghetto Sketches', 1931

Pulp, 'The Trees', 2001

Radiohead, *King of Limbs*, 2011

Robert Schumann, *Jäger auf der Lauer* ('Hunters on the Lookout'), from *Waldszenen* (*Forest Scenes*), Op. 82, no. 2, 1850–51

_____, *Freundliche Landschaft* ('Friendly Landscape'), from *Waldszenen* (*Forest Scenes*), Op. 82, no. 5, 1850–51

Jean Sibelius, 'The Aspen', no. 3, 'The Birch', no. 4, 'The Spruce', no. 5, from Op. 75, 'The Trees', 1914–19

Igor Stravinsky, 'Berceuse', from *The Firebird*, 1910

The Wood

Trad., 'The Trees They Do Grow High'
——, 'The Willow Tree'
The Verve, 'Sonnet', from *Urban Hymns*, 1997
Paul Weller, 'Wild Wood', 1993

Acknowledgements

They helped, root and branch: Julian Alexander, Susanna Wadeson, Lizzy Goudsmit, Deborah Adams, Sophie Christopher, Ella Horne, Nick Hayes, Beci Kelly, Geraldine Ellison, Kate Samano, Josh Benn, Ben Clark, Paula Lester, Mark Hedges, Julian Beach, Freda Lewis-Stempel, Tris Lewis-Stempel, Elizabeth Mitchell, Tracy Pallant, Geoff and Sue Pallant, Leslie Smith, Lucy Hall, David Hill and all the Transworld sales team. And, of course, and most of all, Penny Lewis-Stempel.

John Lewis-Stempel is a writer and farmer. His books include *Meadowland*, which won the Wainwright Prize for Nature Writing in 2015, and *Where Poppies Blow*, which won the Wainwright in 2017. His *The Running Hare* was a *Sunday Times* bestseller and Radio 4 Book of the Week. He is a Magazine Columnist of the Year for his nature notes in *Country Life*. He lives in Herefordshire, where his family have farmed for eight hundred years.

Books by John Lewis-Stempel

England: The Autobiography

The Wild Life: A Year of Living on Wild Food

Six Weeks: The Short and Gallant Life of the British Officer in the First World War

Foraging: The Essential Guide

The War Behind the Wire: The Life, Death and Glory of British Prisoners of War, 1914–18

The Wildlife Garden

Meadowland: The Private Life of an English Field

The Running Hare: The Secret Life of Farmland

Where Poppies Blow: The British Soldier, Nature, the Great War

The Secret Life of the Owl

The Glorious Life of the Oak

Meadowland
The Private Life of an English Field
John Lewis-Stempel

'To stand alone in a field in England and listen to the morning chorus of the birds is to remember why life is precious.'

In exquisite prose John Lewis-Stempel records the passing seasons in an ancient meadow on his farm. His unique and intimate account of the birth, life and death of the flora and fauna – from the pair of ravens who have lived there longer than he has to the minutiae underfoot – is threaded throughout with the history of the field and recalls the literature of other observers of our natural history in a remarkable piece of writing that follows the tradition of Jeffries, Mabey and Deakin.

'Fascinating . . . books have been written about entire countries that contain a less interesting cast of characters'
TOM COX, *OBSERVER*

'Engaging, closely-observed and beautiful'
BEL MOONEY, *DAILY MAIL*

'A rich, interesting book, generously studded with raisins of curious information'
THE TIMES

'A magnificent love letter to the natural world, full of wisdom and experience, written with wit, poetry and love. This is, in fact, one of the best five books I have ever had the privilege to read'
TIM SMIT, THE EDEN PROJECT

The Running Hare
The Secret Life of Farmland
John Lewis-Stempel

The Running Hare is natural history close up and intimate.

It is the closely observed study of the plants and animals that live in and under plough land, from the labouring microbes to the patrolling kestrel above the corn; of field mice in nests woven to crop stems, and the hare now running for his life.

It is a history of the field, which is really the story of our landscape and of us, a people for whom the plough has informed every part of life: our language and religion, our holidays and our food.

And it is the story of a field, once moribund and now transformed.

John Lewis-Stempel writes with insight, wit and poetry. This is a rare and joyful book.

'He describes beautifully the changing of the seasons and the habits of animals such as the hares that make their home in his field. The book is a superb piece of nature writing'
IAN CRITCHLEY, *SUNDAY TIMES*

'Enlightening and stylish . . . Readers who enjoyed the author's last book, *Meadowland: The Private Life of an English Field*, will find much in the same vein here: a mix of agricultural history, rural lore, topographical description and childhood memories. I learned a good deal . . . Lewis-Stempel is a fine stylist, adroitly conjuring scenes in which "medieval mist hangs in the trees" or "frost clenches the ground"'
SARA WHEELER, *OBSERVER*